Dr. John Harris & Jeff Reed

SHARING JESUS ONLINE

Helping Everyday Believers Become Digital and Metaverse Missionaries

Foreword by Dave Ferguson

EXPONENTIAL
Accelerating Multiplication

EXPONENTIAL

Exponential is a growing movement of activists committed to the multiplication of healthy new churches. Exponential Resources spotlights actionable principles, ideas, and solutions for the accelerated multiplication of healthy, reproducing faith communities. For more information, visit exponential.org.

Sharing Jesus Online: Helping Everyday Believers Become Digital and Metaverse Missionaries

ISBN 978-1-62424-112-3 (paperback)
ISBN 978-1-62424-113-0 (epub)

Foreword: Dave Ferguson
Editor: Karen Cain
Cover Design: Sabir Robinson
Interior Design: Karis Pratt

Printed in the United States of America.

Revive.
Recharge.
Refresh.

INSIDE

FOREWORD BY DAVE FERGUSON

SHARING JESUS ONLINE IS A BRILLIANTLY WRITTEN HANDBOOK for digital missionaries, filled with helpful digital tools shared by two courageous digital apostolic pioneers. While there are still many people wondering whether you can do church online or reach people digitally, Jeff Reed and Dr. John Harris are on mission and planting new churches digitally and in the metaverse!

I was excited when Jeff and John told me they were going to write a book to equip digital missionaries and church planters. But I was *thrilled* when they asked me to write this foreword and told me that they wanted to make my B.L.E.S.S. practices the central part of *Sharing Jesus Online*.

One reason for my enthusiasm is my confidence in Jeff Reed's and Dr. John Harris' ability to equip this unique group of Christian leaders. I know this resource will be a great asset because I have seen them successfully launch multiple church planting networks for digital missionaries and planters within NewThing Network.

Another reason for my excitement is that they are translating my book *B.L.E.S.S.: 5 Everyday Ways to Love Your Neighbor and Change the World* (2021) for a completely new audience. The book has already been translated into Spanish and German. Now they are translating it into something new for digital natives! In *Sharing Jesus Online,* Jeff and John have taken one of the best relational evangelistic tools and contextualized it for the online world.

When we (Jon Ferguson and myself) wrote *B.L.E.S.S.*, we did it out of frustration at our inability to share our faith and the love of God with others. I passionately wanted the people around me to have their lives and eternities forever changed by Jesus in the same way he changed mine. But when I tried zealously sharing about my faith with others, people ended up feeling verbally assaulted. When I tried living my life only by example, people simply thought I was a good guy. No one came to know the love of God in Jesus, and I felt frustrated and defeated.

It was in reading the Gospels that we noticed that Jesus earned the nickname "friend of . . . sinners" (Luke 7:34, NIV). Upon closer examination, we discovered that Jesus changed lives by being a friend and intentionally blessing the people who came into his life. There were a handful of practices that Jesus consistently used over and over to bless people. We took those practices and put them into the acronym B.L.E.S.S. so that we and others could easily remember and use them.

It has been amazing to see how God has used these practices, not only in our own lives but beyond! Whole denominations have used the B.L.E.S.S. strategies for national evangelism. Much emerging microchurch training has the B.L.E.S.S. practices at its core. Hundreds and hundreds of churches in the United States and beyond use these practices. Thanks to this book, they can now be used anywhere on the web.

It's always hardest to go first; but thanks to these two forerunners and the content of this book, you won't have to go first or go it alone! You can learn from two of the best as they share their successes and failures in *Sharing Jesus Online*.

PREFACE

"JEFF, YOU REALIZE THAT IF THIS WORKS, it could be bigger than the Catholic Church!" My response to Dave Ferguson's exclamation was that no, in fact, that particular thought had not crossed my mind. I had managed to squeeze in a 15-minute meeting with Dave during an Exponential event based in Chicago in 2021. Ninety minutes later, Dave and I had explored much of my 20+ year history in digital ministry.

Long story short, when I graduated from college in December 1999 I was genuinely shocked that no churches were interested in hiring me to be their digital pastor. In early 2000 (pretty much right after Y2K), I turned unemployment into entrepreneurship, starting my first company: eBeliever.com, an online community of close to 40,000 people from around the world. (Remember, this was several years before Facebook went public, and social media pretty much consisted of MySpace.) In the summer of 2000, I led an online Bible study through the Book of James for 75 people from around the world. My team consisted of family members and college students. Oh, and my main programmer was a junior in high school. Good times.

In the 20 years since, I can tell you some incredible stories of how God is moving in digital and metaverse spaces. But the thing that gets me most excited by far is the evangelistic capacity of a digital missionary. Take it from someone who has been doing digital ministry for literally decades, the way most churches are doing digital ministry will not work, because most of the digital strategies churches are using today are centered around events and broadcasts, not relationships. The key to digital ministry is relationship, which is why the digital missionary holds so much potential.

In recent years, through THECHURCH.DIGITAL and Digital Church Network, we have looked at digital ministry (really the potential digital church) differently. Most physical churches in America use digital tools like social media and communication tools like email and texting to drive people toward physical community in physical church buildings. That's not necessarily a bad strategy, mind you, but we believe that digital resources, and even the metaverse, are more than a communications tool. We believe they are spaces where community happens. And where there's a community, there are opportunities to be the gospel. More than just talking about Jesus, we believe we can *be* the Church in digital and metaverse spaces. While we have seen God do incredible things digitally through "established" churches and organizations, the surprise has been the normal person, literally the person sitting in the physical pew, who gets a calling to share Jesus digitally. The best, most untapped resource the Church has today is sitting in the pews. The opportunity lies before us to start an evangelistic movement of global proportions in digital and metaverse spaces.

The key to digital ministry is relationship, which is why the digital missionary holds so much potential.

Honestly, as I was recounting some of my stories for Dave Ferguson and sharing my vision for the digital missionary, Dave was probably more excited about the idea than I was!

In subsequent meetings with Dave and the team at NewThing Network, we discussed the necessity of creating a disciple and growing that person's leadership ability. We explored unique expressions of the Church operating in digital and metaverse spaces. We talked about what it would look like for those churches to multiply into networks for connecting friends on mission. But as we set out to start this digital missionary movement, it became obvious that we were missing a crucial step. We were missing the framework for our digital missionaries!

It wasn't long before the NewThing team brought up Dave and Jon Ferguson's B.L.E.S.S. practices. You may know this framework: Begin with prayer, Listen, Eat together, Serve, and Share your story and Jesus' story. At the time, I was vaguely familiar with the system, but I knew immediately that the whole "eat together" thing wasn't going to work out digitally. Dave and Jon supported us in adapting their ideas, even allowing us the freedom to contextualize their established framework to work in digital or metaverse spaces. They were so supportive! Honestly, I'm pretty sure it was Dave's idea for us to write this book!

Early on, it was obvious I needed to invite others into the conversation. To contextualize B.L.E.S.S. for digital and metaverse missionaries, I (Jeff) had conversations with pastors, planters, and practitioners of digital and metaverse churches. One individual's passion quickly stood out: Dr. John Harris, leader of a digital discipleship community called Hearts Like His. As lead pastor of Hearts Like His, Dr. John has a huge heart for reaching and discipling people in the digital space. Let me have Dr. John share a little bit of his story, and his passion for digital evangelism.

• • •

To reach the 65 million people who have left the Church[1] and to share a more gospel-centered Christianity than many have experienced, my wife and I (John) started Hearts Like His as a Facebook group. I had previous success building physical communities such as small groups, but I wondered whether it was possible to do it online. Our hearts are for discipleship, but more specifically transformational discipleship. We wanted to see people's lives transformed. So we were not just aiming for community. We were aiming for transformation in a discipleship community. So we wondered, Can we really develop authentic relationships and a sense of community online? Could we really see people grow in their relationship with God through screens?

We discovered that the answer was a resounding yes! In some cases, people were experiencing more community online than in person. One woman who had been a member of her church for years, actually said, "Being part of Hearts Like His has shown me what I was missing from my in-person church." As group members invited their friends and family members, one person was hospitalized for a few days due to surgery. He told his nurse about Hearts Like His, and she joined.

The growth and success of Hearts Like His drew Jeff's attention, and we started to connect about what was happening. That resulted in a couple of podcast interviews about our approach to digital ministry, specifically our goal of reaching the de-churched and the unchurched with a focus on the Father's heart as expressed in the gospel message.

When Jeff and I first met to talk about digital missionaries and B.L.E.S.S., I jumped at the idea of writing this book. The opportunity the Church has to impact people online for the glory of God is immeasurable. It's not hard to see how the online community has the largest unreached people group. The potential in the digital mission field coupled with the tools technology has given us to reach the world with the gospel is an amazing combination.

• • •

In the coming pages, you'll discover a simple framework that anyone can use to share Jesus digitally. This does not require massive technological prowess, nor does it require millions of followers. Everyone has some influence digitally that can either be used to connect people to Jesus, or it can be wasted.

Introduction

THE B.L.E.S.S. FRAMEWORK HAS PROVEN to be an effective evangelism strategy based on developing relationships with people that can be really effective in digital and metaverse environments. "The B.L.E.S.S. practices were created as a simple tool to help you bring the love of God to others."[2]

How can you be part of what God is doing around the world online? I (John) want to begin this book by answering three leading questions: Why now? Why you? Why B.L.E.S.S.?

Why Now?

People Are Searching for More Answers

"In October 2022 there were 4.74 billion social media users worldwide, equating to 59.3% of the global population. Social media numbers have continued to grow over the past 12 months, with 190 million new users joining social media since this time last year."[3] Not only are these people online, but they are searching online! They're Googling for answers, for comfort, for truth, for healing, for hope—for everything you and I know can only be found in Jesus. Check out these search results from 2022.[4] This is a list of top spiritual or religious questions people are typing in the Google search engine, along with the number of searches *every month*.

What is love?	7,480,000
Who is Jesus?	2,740,000
What is the Big Bang?	2,740,000
What is the church?	2,240,000
What is money?	1,830,000
What is sin?	1,500,000
Who is God?	1,500,000
What is life?	1,220,000
What is the Bible?	1,000,000
What is hell?	1,000,000
Where are the churches near me?	823,000
What is prayer?	823,000
What is evolution?	673,000
What is a Christian?	673,000
What is a prayer for healing?	450,000

This is what the idea of "sheep without a shepherd" (Matthew 9:36, ESV) looks like digitally. These numbers represent questions people are asking every month! Now, you might ask, "Why don't they just come to church and get answers to their questions?" Well, here's why.

Trust in Institutions Has Decreased

One major cultural shift has been a general decrease in trust of institutions. The Church is one of these institutions. "Gallup's survey of confidence levels in institutions found significant declines for 11 of the 16 institutions." In 1973, 65% of Americans said they had "quite a lot" or "a great deal" of confidence in the Church or organized religion. That number as of 2022 is now 31%.[5] This is an external factor, one that is outside of our control. With a decrease in institutional trust, we may not want to keep putting our institutional foot forward first. The dramatic decrease in trust in most institutions has a domino effect on the Church and must therefore impact our methods of sharing the gospel message. That's the "Why now?" So let's talk about "Why you?"

Why You?

The Church Has a Credibility Crisis

Culture has shifted, and the Church is no longer considered a "good" or "safe" place. In addition to cultural changes outside the Church, cultural views toward the Church are changing due to internal problems. In addition to the 65 million Christians who have left the Church in recent decades, Christian influencers who have left Christianity have popularized the deconstruction movement, and numerous pastoral scandals of infidelity, child abuse, and sexual assault have decreased our credibility. And let's face it, we were not our best selves during the racial and political division of 2020–2022. I'm not saying that the Church is supposed to be buddy-buddy with the culture, but we at least need enough credibility to share the message of Jesus.

Trust in Pastors Has Decreased

When trust in the Church drops, so does trust in her pastors. In 1985, 67% of Americans said pastors had a "high" or "very high" ethical standard and level of honesty. That decreased to 40% in 2018 and further to 37% in 2021.[6] The days are long gone when inviting people to hear your pastor's sermon was the primary step. There are plenty of gifted pastors who work hard every week to deliver God-inspired, biblically sound sermons, but you are going to meet people who do not want to hear them. Pastors are not seen as sources of spiritual authority or even moral authority as much as they were years ago.

Church Invitations Are Not as Effective

When I (John) served as an executive pastor, I wished more Christians would invite their family, friends, co-workers, and neighbors to church. Church invitations have been a primary means to grow churches for many years, especially in America. Inviting non-Christians to a weekly service or church event is often called "invitational evangelism." That approach will still work, but only for a shrinking number of people. A Barna study from 2020 found that "about half of all unchurched adults

(52%) would not be interested in any invitation, to any church or faith environment."[7] This statistic applies to physical, digital, and metaverse. I'm not suggesting we stop inviting people. I'm saying that the primary means of reaching people can no longer be trying to get them into the sanctuary. There is a growing majority of people who are not coming no matter what we do to try to get them there. So the answer is not trying to figure out how to get better at inviting but instead to get better at equipping and sending.

I share these statistics not to paint a negative picture but to educate you, since most Christians don't follow Church trends. If we do not understand the cultural context we live in and how we are being perceived, then we will operate from a cultural blind spot that will sabotage our efforts and subsequently our effectiveness. Your pastor has likely been aware of these statistics, but I feel like you need to know so that you, the everyday believer, are clear on why *now* is the time for you to become a digital missionary. Knowing how you are perceived by the people you want to reach is critical for relationship building and overcoming obstacles to mission effectiveness. These are current realities you need to know about. Many of you reading this book have depended on your church or pastor to do the work of evangelism. Well, those days are gone. Now you, the everyday believer, must carry the torch so that people who will never come to church can hear and receive the gospel.

Yelp for Jesus?

We have all had bad experiences as customers. In order to avoid them as much as possible, it has become second nature to look at customer reviews online before making a purchase. Yelp reviews made this popular. It gave customers and clients a platform to share their experience of a company's products and services. When it comes to spending money, customers trust the feedback of another customer they do not know over what a business says about its product or service. Think about the impact Yelp reviews have had on our culture. What the business says doesn't matter as much anymore. Customers want to hear from other customers. They want to know your experience. It's the same way with the Church. People will trust what you as a "customer" say about Jesus more than what the Church or pastor says. It's not that people do not trust anyone at all. But they have changed who they trust and how much they trust, at least at first. So let's recognize that and lean into it. Evangelism today is sharing with people your "customer review" of Jesus. So, yes, you need to be ready to give a Yelp review for Jesus. Tell people how he changed your life. The good news is that you can play a significant role in turning the tide of credibility and trust by sharing the gospel of Jesus in your social circles in person and digitally. Remember, you are the solution! And this book is a tool to help you.

Why B.L.E.S.S.?

It has been said that people don't care how much you know until they know how much you care. In the context of evangelism, that means they don't care about the gospel you want to share with them until they know you genuinely care about *them*. This is why relational evangelism is so effective. Relational evangelism, also called friendship evangelism, focuses on creating trusted relationships with people to reach them for Christ.

> Anywhere relationships exist, evangelism should too.

The B.L.E.S.S. framework is a relational evangelism model that has been so effective for in-person relationships that Jeff and I (John) encourage you to use it in digital and metaverse environments. After all, relationships are relationships—and anywhere relationships exist, evangelism should too.

At a high level, to bless someone, we Begin with Prayer, Listen, Enjoy Shared Experiences, Serve, and Share Your Jesus Story. Even though this framework appears quite simple (and it is) here's a quick overview of the digital version of B.L.E.S.S.:

B: Begin with Prayer

I (Jeff) love this idea of beginning in prayer. Confession time here. On the Mary/Martha spectrum, I am 100% Martha. My friends would agree! I'm a recovering workaholic and am guilty of living my spiritual life more in line with "Ready, fire, aim!" instead of "Ready, aim, fire!" If we actually let the Holy Spirit lead us, would things be different? What if we operated on God's timing instead of us forcing ours?

"Likewise the Spirit helps us in our weakness. For we do not know what to pray for as we ought, but the Spirit himself intercedes for us with groanings too deep for words. And he who searches hearts knows what is the mind of the Spirit, because the Spirit intercedes for the saints according to the will of God" (Romans 8:26-27, ESV).

This is why, from a practical standpoint, I find "Begin with Prayer" to be such a good reminder. Prayer comes as naturally as breathing for some people. Others of us need a reminder to be intentional with what we're praying. As Romans 8 says, we often do not know how to pray and lean on the Spirit to guide us as we intercede for ourselves and others.

L: Listen

In 1995, Billy Graham preached a sermon that was heard, in real time, by 20% of the world's population.[8] Well, Toto, we're not in 1995 anymore. People are not responding to sermons and lectures as they once did. Instead, digital and metaverse missionaries like us must listen before we talk! We should ask questions rather than make statements—listen to the person talk, and learn what the Holy Spirit says in the situation at hand.

"My dear brothers and sisters, take note of this: Everyone should be quick to listen, slow to speak, and slow to become angry" (James 1:19, NIV).

In digital and virtual reality space, posture is everything. Listening to someone you don't know will ultimately give you credibility to speak. Soapbox preachers don't work digitally. Digital and metaverse missionaries need relational credibility, and one of the fastest ways to get it is to listen.

E: Enjoy Shared Experiences

In the original B.L.E.S.S. practice, the E was "Eating together." So my first question was obviously, "Dave, how in the world are we going to eat together digitally?!" I (Jeff) mean, I guess we could eat pixelated food in virtual reality, but that's so limiting . . . and maybe a little weird. So the next question was how to replicate this physical experience in digital and metaverse environments.

In May 2022,[9] Digital Church Network brought in 15 digital and metaverse pastors and planters, hosting a conversation around what the E could stand for. Honestly, I came into that discussion with no ideas or expectations. More than anything, I was curious to see how the Lord would lead us.

"So I recommend having fun, because there is nothing better for people in this world than to eat, drink, and enjoy life. That way they will experience happiness along with all the hard work God gives them under the sun" (Ecclesiastes 8:15, NLT).

Through this roundtable discussion with pastors and planters,[10] we discovered that the key to relationally connecting with people in digital and metaverse spaces is having a shared experience. Ecclesiastes talks about eating, drinking, and enjoying life. The Bible literally tells us to have fun! Rather than eating physically, what if we utilized experiences that people enjoy digitally or in the metaverse to build relationships? Digital missionaries have the opportunity to have fun!

So what can a digital or metaverse experience look like? How does a physical shared experience differ from a digital or virtual reality experience? As a digital missionary, this should be the most fun element for you, as it allows you to bring your personality into play.

S: Serve

When you say "serve" to a bunch of Christians who attend physical church services, they immediately reflect on serving at the physical church building. Maybe they're greeters or workers in children's ministry or on the tech team. There's nothing wrong with this context, but digital missionaries must do something different.

Continuing in the posture of humility, how can we serve the person God is pointing us toward? What would it mean to your relationship if you took the time to serve them, to help them on a more intimate level, to really understand humility?

"In everything I did, I showed you that by this kind of hard work we must help the weak, remembering the words the Lord Jesus himself said: 'It is more blessed to give than to receive'" (Acts 20:35, NIV).

As digital missionaries, this is another fun category to work with. How do you serve someone digitally? What does serving someone look like in the metaverse? Some imagination and creativity will be necessary to make this happen, but rest assured, there *are* ways to serve people digitally.

• • •

Let's stop for a second and review our journey thus far.

In an attempt to attune our heart to God's, we **Begin by Praying** for an individual. As we feel the Spirit leading, we start a conversation with the person not by pushing an agenda but by **Listening** to what they have to say. By truly listening, we learn more about the person, what their hopes and dreams are, or maybe something as simple as what they like to do for fun. To strengthen the relationship, we **Enjoy Shared Experiences** and get to know them better, which leads us to a place where we can **Serve** at an individual level. Led through prayer attuned to the Holy Spirit, this person will feel heard, connected to, and cared for. All this pre-evangelism leads to the place where you can now **Share Your Jesus Story**.

S: Share Your Jesus Story

Think about the relational equity we have invested during this process thus far. Through praying, listening, enjoying shared experiences, and serving, we as digital missionaries are building relationships that will establish trust. Culturally (whether good or bad), evangelism has moved from a rational decision to an emotional one. The credibility earned as we B.L.E.S.S. other people allows us to connect with them on an emotional level.

"But you will receive power when the Holy Spirit has come upon you, and you will be my witnesses in Jerusalem and in all Judea and Samaria, and to the end of the earth" (Acts 1:8, ESV).

It's a Framework, Not a Formula

As we move through the steps of this framework, I (John) want to highlight important principles to keep in mind. We call this a *framework* instead of a *formula*. With a formula, like a recipe, you have to follow the steps sequentially and get each one just right for the process to work. Think of this framework as a group of principles with freedom and flexibility. If you want to follow it precisely as written, you can; but know that you have the flexibility to move the steps around based on how you are led by the Spirit and how your conversations and relationships develop. Evangelism is a spiritual process and does not always follow the same sequence because the Spirit leads differently in different situations. It would be a missed opportunity if you met someone and after a short conversation they were ready to hear the gospel but you delayed because you thought you had to have a shared experience first. One step does not always have to follow another in order.

Also, the Spirit may move in the conversation or relationship in a completely different way. The purpose of this framework is not to tie you down to a method but to answer the questions "Where do I start?" and "How can I build relationships so that I can share the gospel at some point?" You may want to use some of these principles in current relationships, or you can definitely use these to move someone from a complete stranger to a friend. B.L.E.S.S. is a framework, not a formula; an opportunity, not an obligation; a set of principles, not prerequisites. Let's dig deeper and see what B.L.E.S.S. can look like in digital and metaverse spaces.

What Does *That* Mean? Defining New Terms

Throughout this book, we will introduce new ideas and perspectives on using the digital community for the Kingdom. In doing so, we may throw some new words at you, or we may have new definitions for common words as we're trying to set a new standard for the Kingdom. Let's explore some of them:

DIGITAL: Commonly referred to as Web 2.0 technology, *digital* refers to social media, mobile phones, streaming, and Zoom. Digital encompasses many of the ways we use technology to communicate, except for metaverse technologies (including virtual reality).

DIGITAL MISSIONARY: A digital missionary is an everyday person (as in not a pastor or church leader) who is sharing Jesus with people who are "cold to Christ." There's a movement where these missionaries are now functioning outside the established Church. A digital missionary has been sent by Jesus to share his message of salvation with the world, and they carry out this mission in digital and metaverse spaces. One thing that characterizes a missionary mindset is living a sent lifestyle. A digital missionary will use digital relationships and community to B.L.E.S.S. people, sharing Jesus' love with them.

DIGITAL EVANGELISM: Digital evangelism is the intentional and strategic use of internet platforms to share the message of Jesus with people using digital technology. Digital communities are the new mission field. Digital evangelism is a newer method to reach them.

METAVERSE: The metaverse is commonly referred to as Web 3.0 technology. (We are not specifically referring to Facebook's "Meta," although it's included.) The metaverse comprises a suite of technologies including virtual reality, augmented reality, mixed reality, blockchain, cryptocurrencies, artificial intelligence, etc. Some of these technologies will be readily available in 2023; some are not yet released. Still, the metaverse is having an immense impact on today's culture and community. The metaverse is commonly referred to as "the Age of Experiences."

VIRTUAL REALITY: Arguably the most popular metaverse technology today, virtual reality involves putting on goggles that block your view of physical reality and instead take you into a computer-drawn "virtual" reality. Sometimes mistaken for a video game, virtual reality is actually quite real, and it's an area where authentic relationships develop in computer-generated environments.

COMMUNITY: At its basic level, a community is a group of people who share an identity and a mutual concern for one another's welfare. So when we talk about online communities, this is the idea we have in mind. The Church has historically defined community as existing only in physical space. As digital and metaverse technology expands, it offers new opportunities for evangelism, discipleship, and even church planting.

CHURCH: Simply put, we define church as a group of believers who gather regularly to worship God, to help each other grow in Christ, and to engage the world with the gospel. As we examine digital and virtual reality churches, there are many similarities to the physical church and also some differences. In many ways, advances in digital or metaverse technologies have allowed for the church to be re-imagined, not moving away from Scriptural ecclesiology but instead re thinking it and mapping those functions across the technology. In many ways, digital and metaverse churches can more closely align to biblical ecclesiology than the Western physical church model (for example, the everyday access people have to each other). More information on this can be found in my (Jeff's) book *VR & the Metaverse Church: How God Is Moving in This Virtual, Yet Quite Real, Reality*, published through Leadership Network. Find more information at http://leadnet.org/vrbook.

Who Is this Book For?

This book is for every believer who wants to establish relationships with people to reach them with the gospel of Jesus. This can refer to relationships with unsaved people you already know or to new relationships you want to start. The B.L.E.S.S. framework responds to the question, "How can I build relationships so that I can share my faith online? I would do it more if I knew what to say, when to say it, and who to say it to." If this is you, we are glad you picked up this book.

> We define church as a group of believers who gather regularly to worship God, to help each other grow in Christ, and to engage the world with the gospel.

Its message is for all believers, no matter your denomination or level of faith or Bible knowledge. Whether you are a new believer or you have been saved for many years, these principles will be helpful.

Sharing Jesus Online is not a church leadership book. It is written to the everyday believer . . . the person in the pew. There will be a massive cultural shift as a result of metaverse technologies in the coming years. This shift will move to empower individuals, not organizations. We as individuals have the opportunity to B.L.E.S.S. people who the building and the pulpit aren't reaching. Sounds almost biblical, doesn't it? Acts 8:4 (NIV) says that "those who had been scattered preached the word wherever they went." What if we used the internet to scatter the gospel as the Church scattered long ago across those Roman roads? What if the gospel spread in digital and virtual communities, reaching new people as it did in ancient cities back in the time of Acts? We believe it can—and we want to help.

Sharing Jesus Online (the Course)

In addition to this book, John and I (Jeff) are developing an online course that can help you and your church take intentional steps toward becoming digital missionaries using the B.L.E.S.S. practices. Along with the course, you'll find a digital community to talk with and encourage you on your journey. For more information about *Sharing Jesus Online,* visit http://digitalchurch.network/missionary.

CHAPTER 1

Reimagining the Great Commission

Aslan Is on the Move!

ONE OF MY (JOHN) FAVORITE LINES comes from the movie *The Chronicles of Narnia: The Lion, The Witch, and The Wardrobe,* based on the book series by C. S. Lewis. Peter, Susan, Edmund, and Lucy have stumbled through the back of the wardrobe into Narnia. As they are trying to figure out what happened and where they are, they meet the Beaver. He brings them to his home for dinner. As Beaver and Mrs. Beaver are giving the children an orientation about Narnia and letting them know about the dangers of the White Witch, Mrs. Beaver says, "But there is hope, dear." Turning toward Beaver she asks, "Want some hope?" Beaver replies, "There's quite a bit more than hope."[11] He leans over the dinner table to whisper, "Aslan is on the move!" (Aslan is a lion that represents Jesus Christ.) If you read the book, it adds this detail: "And now a very curious thing happened. None of the children knew who Aslan was any more than you do, but the moment the Beaver had spoken these words, everyone felt quite different."[12]

I want to start *this* book with Beaver's words: "Aslan is on the move!" *Jesus* is on the move, changing lives in digital and metaverse spaces. Here are a handful of examples:

In 2017, Angela Craig launched the first online church in the Assemblies of God on Facebook. It grew to 14,000 members reaching 45 countries and 43 languages. I remember her telling me, "I've never met him in person, but our worship leader lives in Uganda."

When over 700 new churches went live for the first time on the Church Online Streaming Platform, 430,000 people gave their lives to Jesus.[13]

The Luis Palau Association has an online evangelism campaign called Hope with God. In 2021, they presented the gospel to 6.2 million people and recorded 1.3 million decisions for Christ.

A 2021 Barna report surveyed 8,000 Americans and found that 54% of them are open to learning more about Jesus, especially those with children.

Endale is a digital missionary to people mainly living in Central Asian and Arabic countries. He started organizing WhatsApp groups for about 20 people at a time, recording his voice and sending it to people in the group as they listened and reflected. He has trained over 300 digital missionaries who share the gospel in one-on-one conversations. Together they have reached more than 3,000 people, with nearly 100 of them receiving Christ and going through a discipleship process in WhatsApp. The 300 members of his church don't have a way to safely meet together for fellowship. The WhatsApp church is the only way they can grow in their faith and do ministry.[14]

Even miracles are happening through digital ministry. Church Anywhere is a digital church extending from First Capital Christian Church in Corydon, Indiana. One of their online pastors, Megan Carter, shared this account with us:

STORY

In October of 2020, I posted a video on TikTok of me writing on the whiteboard in my office all of the names of the people I had prayed for on TikTok. Erica commented on that video, "Girl, you prayed for my family and I as we were told our unborn baby had trisomy 18 and that he wouldn't survive. At our

last appointment we were told he is totally fine. Healthy and healed! No more brain cyst! No more holes in his heart! No more clubbed foot! He's chunky and growing healthy. Praise God!!!!"
God is moving in the lives of people, even when we've never met face-to-face. If we are willing to use the tools he has given us to reach people right where they are, God will step into the life of the person on the other side of the screen. You may not always hear stories like these from every encounter with someone, but that doesn't mean that God's not working in their life.

Hold on! Did Erica say that her unborn baby was healed because of a TikTok prayer? Yes, I think she did! C'mon, Jesus!

But wait! There's more! Articles like these keep popping up as more digital missionaries are taking the gospel "as they go" online:

» The Gospel Coalition published "Innovative Evangelism: Bringing the Gospel to TikTok."[15]
» Medium.com published "God Mode Activated: Meet the Gamers Bringing Jesus to Twitch: Christian Evangelists Use Streaming Platforms to Spread the Gospel While Playing Fortnite" (quoting and referencing many of Jeff's friends in gaming churches).[16]

More updates like these are turning up:

STORY

I've spent nearly 30 years traveling the world preaching Christ . . . but there has not been a single day [in recent memory] where I've not seen decisions for Christ on TikTok. I wake up every day to comments like this: "You don't know me, but I prayed to become a Christian months ago on your account. I want you to know that I have my first Bible and will be getting baptized next week.[17]

People all over the world are getting saved through screens! It's already happening! Not only is Aslan on the move, but he is moving online through digital missionaries who have

> People all over the world are getting saved through screens!

reimagined the Great Commission, "Go and make disciples of all nations" (Matthew 28:19, NIV).

Revisiting the Great Commission

In recent years, evangelism has taken a beating. Whether looking at the role of the "evangelical" in politics or some historical missteps that continue to sustain tension, the Church has lost sight of what evangelism could, albeit should, be.

Here's the Church . . . Where Are the People?

In recent years, the Church has found herself in an echo chamber of sorts, surrounded by people who sound like her, act like her, and think like her. Unfortunately, many people have walked away from the church this season, with an estimated 23% decline in US church involvement over the past 20 years.[18] The "US church model" has struggled for many reasons (an excellent focus for another book), but we want to focus on the fact that today's church has an opportunity to think differently. As the Church looks to revive the Great Commission, she must revisit "church" through the lens of digital and metaverse opportunities.

> Here's the church.
> Here's the steeple.
> Open the doors.
> Where are the people?

In the modern era, the Great Commission's directive to "Go and make disciples of all nations" sits differently. The largest nation in the world is Facebook, if we look at social media networks as "nations." There are more active Facebook users than inhabitants of China and India combined. Some churches, alongside some pioneering individuals, are working diligently to bring Christ into the digital or the metaverse world. Truthfully, the missiological implications in this space are huge. We need to not only revisit the Great Commission and get back to our missional roots, but we need to reimagine what the Great Commission looks like with the technological tools we have available.

Digital Missionaries Reimagining the Great Commission

Digital Opportunities to "Go into All the World"

In addition to "Go and make disciples of all nations" in Matthew 28:19, Mark 16:15 records another iteration of a commission when Jesus said, "Go into all the world and proclaim the gospel to the whole creation" (ESV). What I (John) love about these commissions is the strong emphasis on "Go" (or perhaps more accurately translated "as you go"). Digital opportunities and capabilities change what it means to "Go into all the world" (Matthew 28:19) because of how technology enables us to connect to the world. Jesus said, "Go" because based on the options available at the time, his disciples had to *leave where they were* to contact new people in other places and communicate the gospel to them. Now we can do that with smartphones and laptops. The tap of a finger or the click of the mouse immediately connects us with people in "all the world." We can reach the nations without having a passport or even getting off our couches.

Digital evangelism allows us to bring Jesus to people through phones, laptops, and goggles right where they are–anywhere around the world–for free!

Technology has given us a global audience, and therefore the world is at our fingertips. We can have an immeasurable impact online if we become more intentional. We should be ready and able to share the gospel in any place where we spend time with people. Online platforms provide the perfect place and the perfect opportunity for evangelism. Digital evangelism allows us to cross national borders and eliminate geographic distance to bring Jesus to people through phones, laptops, and goggles right where they are—anywhere around the world! And in most cases, we can do it for free!

York Moore, president of the Coalition for Christian Outreach, shared this story:

STORY

God is using TikTok in a powerful way to reach Gen Z with the gospel. Gen Z is *not* tuning into Christian radio or satellite TV; [their] eyes and ears right now are undoubtedly on TikTok, and so this is where we bring the unction of the Spirit of the Living God. We may not be able to pray in schools, but every single day I get to speak 6-12 inches away from the eyes of a 14-year-old sitting in the backseat of his mother's SUV after school and tell him that Jesus loves him and that he can be born again through repentance and faith. There is no greater honor or joy for an evangelist![19]

We have digital opportunities and digital capabilities, but in order for us to take advantage of these, we need a digital mentality. This will come as we make some necessary shifts in how we "do" evangelism.

Metaverse Opportunities to Reach the Unreachable

Throughout this book, we will be taking care to separate the digital and metaverse mission fields. We do so for two reasons:

1. At the time of writing, metaverse technology is still very young, and in many ways it has not yet been adopted globally. As a result, we see value in separating this technology from the more accepted "digital."
2. Because it is not yet accepted globally, the metaverse mission field operates much differently than the digital mission field.

As of 2021, there are 57.4 million virtual reality users in the United States.[20] Of the 331.9 million people living in the US at a given time,[21] roughly 17% of them utilize virtual reality. Because virtual reality has not yet received global adoption at the time of this writing, the chances of two virtual reality users knowing each other are rather slim. But the opportunities virtual reality provides to have conversations with people who are open to Jesus are phenomenal! Still, in 2023 the mindset of the metaverse mission field is different from the digital one. Your digital network is likely filled with people you can reconnect with, while your virtual network likely shows the potential of who you can reach. Both networks are valid and necessary if we are to fulfill the Great Commission.

A digital missionary has an opportunity to reach beyond the church walls, whether in physical, digital, or even metaverse space. I (Jeff) talked about churches in virtual reality in my recent book entitled *VR & the Metaverse Church: How God Is Moving in This Virtual, Yet Quite Real, Reality* (Leadership Network, 2022). Whatever the environment, even virtual reality, God is moving and people are coming to Christ through churches. There's an opportunity to connect with people on an emotional level and introduce them to Jesus in physical, digital, and virtual spaces.

Churches that embrace digital and metaverse communities and plant digital and metaverse churches in those spaces will effectively reach different types of people. For example, Jason Sprowl, associate editor for *Outreach* magazine, pointed to this when he said, "We'd be hard-pressed to find a higher concentration of unchurched or dechurched people from the 18 to 35 demographic than we can find in the metaverse."[22] Hopefully you are seeing patterns arise and are beginning to recognize what many Christians worldwide have discovered: it is possible to share Jesus and disciple people digitally and in the metaverse. Not only is it *possible*, but it is *necessary*. You are necessary!

Re-Engaging the Great Commission

Necessary Shifts for Re-Engaging the Great Commission

As a believer, you must start seeing yourself on the field running plays instead of in the stands cheering for others. On a regular basis, you have digital interactions with people who need to know Jesus. They're in your newsfeeds, live chats and streams, virtual reality, and many other places. Beyond being equipped to be a digital missionary, I (John) want you to see the necessity and urgency of you *becoming* one. In light of what I've shared about the culture and the Church, I see the following shifts as necessary. The purpose of this book is to be part of that process.

Shift #1: From Your Pastor's Sermons to Your Personal Story

Your pastor has a calling and so do you. As God moves in the hearts of those people you interact with online, make an intentional effort to start seeing them as people God sent to you—not people God sent for you to refer to your pastor. This is *your* moment, and God wants that person to hear what he has done in *your* life, not your pastor's. After seeing the statistics mentioned earlier, you can understand why this is true. People are more interested in hearing about your experience with God than in hearing a stranger's sermon about God. (We will discuss this in more detail later.)

God sets up what I call "God moments," the moments when there is a divinely orchestrated interaction with someone else. And, yes, it is a "setup!" God intentionally set you up with the conversation and relationship. You've invested into this person, and if you've reached the point of sharing your story, that is *your* moment. You need to know how to do more than invite them to church.

Shift #2: From Talking About the Church First to Talking About Jesus First

Every believer is called to be a witness about Jesus, but we have become better witnesses for our churches. "I've been praying for my friend to come to church for a long time" or some variation of this is a common statement made by Christians who believe that inviting someone to church is the first step. What if you knew a different method? What if you invited someone into the Kingdom before you invited them to your church?

Inviting people to church is a popular way to get a church to grow, and there's nothing wrong with it. But in light of our cultural realities and biblical mandates, you should be equipped and ready to talk about Jesus first. People should not have to come to your church to hear about Jesus. It's so ingrained in us, however, to just talk about church. We have to shift to talking about *Jesus* first because the amount of people who are never coming to your church (or any other) is rising. Christians could be sharing Jesus with many more people if we made this shift.

Shift #3: From Focusing on a Weekly Event to Everyday Interactions

There is so much work that goes into weekend services that we tend to look at Sunday as the Super Bowl and the rest of the week as halftime. This needs to change! We need to shift our mentality to seeing Sunday as halftime, when we meet as a team to get inspired, coached, and encouraged. Then what happens during the rest of the week is the Super Bowl, when we go out and put our strategies into action—including sharing Jesus online. How else are we going to get back to the days when "the Lord added to their number day by day those who were being saved" (Acts 2:47, ESV)?

Shift #4: From Casual Scrolling to Intentional Seeking

Think about how many hours a week you spend online or specifically on social media. The average adult spends 2 hours and 27 minutes a day on social media.[23] That is 17 hours and 15 minutes a week. What if you took just one of those hours and made it intentionally missional? We're not talking about adding something else to your already busy schedule. We're talking about adding meaning and Kingdom purpose to something you are doing anyway. Jesus came "to seek and to save the lost" (Luke 19:10, ESV). Now he wants to do that through *you* online—which leads to the next shift.

> What if you took just one of those hours spent online and made it intentionally missional?

Shift #5: From the Ordained to the Ordinary

Pastors are some of the hardest working people in our nation. Traditionally, churches have depended on the pastors to do all or most of the work. More and more pastors and churches are shifting their focus "to equip the saints for the work of ministry" (Ephesians 4:12, ESV). It's common to expect the small number of clergy at a church to do all the ministry, but that ideology has to shift. We as ordinary believers need to learn when and how to engage people directly. God is sending out everyone! Each of us has been sent by Christ to be his ambassador. It's time for you to carry the gospel with you and lead people to Jesus where you live, learn, work, and play. That's where you will see God move in the lives of those around you.

Digital Relationships and Communities

Since the beginning, Christians have used technological advances to advance the gospel, from the Roman road to the printing press to the radio, TV, and beyond.

The building blocks of digital communities are digital relationships. Do digital and metaverse relationships replace in-person relationships? Absolutely not! But there are no limits to the levels of connection and intimacy that can be created in digital spaces, especially with the tools to do so.

Digital Community

The mission of Facebook is "to give people the power to build community and bring the world closer together."[24] It has done just that. It's likely that you are in a few Facebook groups. You have ongoing conversations with people you have not met in person and probably never will. You ask questions and get answers or answer the questions of others, and you get support and give support in Facebook groups that align with your interests. Why? Facebook has given the world the technology, platform, and ability to build community. Although Facebook is the largest social media platform, many of you are already experiencing authentic connections and genuine digital relationships on other platforms as well. It's the norm in our culture now.

During a pre-conference session about digital ministry at Exponential 2022 in Orlando, I (John) once asked a room full of pastors and church leaders, "What are the elements that create a sense of community? What do you think people are looking for?" Immediately, hands were raised all across the room. As I pointed to the raised hands, people were yelling out words such as *transparency, safe place, belonging, support, encouragement, being seen, being cared for,* and others. Then I asked, "Why is it that no one said, *in-person*?" You should have seen the light bulbs go on all across the room. I said, "Because that's not a requirement. You can be in person and if you do not feel a sense of belonging, support, safety, etc., you will still not have community."

The world has changed, and in the process culture is redefining the meaning of gathering, what it means to connect, and what is necessary to build quality long-lasting relationships. These changes affect everyone in the world and must be considered in our approaches to ministry and accomplishing the mission of Christ. In the New Testament, there was only one way to gather, and that was in person. Due to the advancement of technology, the idea of gathering has been expanded so it can be done digitally . . . even virtually.

Culture is redefining the meaning of gathering, what it means to connect.

The whole world has embraced the significance, value, and power of digital relationships and community building, but when it comes to matters of evangelism and discipleship, some in the Church still argue over their legitimacy. If you're reading this book you probably already get it. You might be thinking, "I'm already on board! Just show me how to do it!" This book was written as a blueprint to help you take intentional steps toward building relationships with those who do not know Jesus. The B.L.E.S.S. framework can also help you reach those who have left Christianity (called the "Dones" for being done with church). Maybe you'll be on the front lines, changing the "Rise of the Dones" to the "Return of the Dones."

CHAPTER 2

Overcoming Obstacles and Misconceptions

NOTE: This may be the most important chapter in this book.

My Learning Curve

AS A FULL-TIME PASTOR, I (JOHN) REMEMBER becoming aware that my social circle consisted only of Christians. From church staff to small groups, there were no opportunities to share the gospel in those circles. I had to be intentional in conversations with servers at restaurants or other places I would hang out. I needed to learn how to develop relationships with unsaved people that went beyond small talk.

I also had to learn how to become more comfortable talking with strangers about Jesus. This might sound odd to you because I'm a pastor—you may assume I already knew how to do that, but I didn't. Most of my experience sharing the gospel these past 30 years took place on a stage behind a microphone. I did all the talking, there were no questions, and someone else followed up with whoever responded. Strange to think about, huh? In this book, I'll share how I learned to get better at initiating conversations with strangers and becoming more comfortable talking to them about Jesus. The more I learned and overcame obstacles I'd been unaware of, then the more comfortable I was having faith conversations with people in person and online.

Some of you have a passion for evangelism that wants to break forth like a free-flowing river, but it is trapped behind a dam of fear. However, the fact that you're reading this book means the dam of fear is breaking down and some water is trickling out. I (John) have learned that one of the most important things a leader can do for a team is to remove obstacles to the team's success.

The following obstacles and misconceptions can act as a dam blocking the river of your evangelistic ability, passion, creativity and boldness. One Barna report showed that 56% of practicing Christians have had two or fewer conversations about faith with a non-Christian in the past year.[25] The same report, however, showed that "those who had at least one conversation about faith came away more confident and eager to talk with others."[26] In this chapter, we want to identify a few obstacles and misconceptions that can keep you from realizing your missional potential. When I think about obstacles to evangelism, I cannot help but begin with my own experiences and feelings about evangelism. So I (John) will share the obstacles, while Jeff addresses the misconceptions. As we share our thoughts, see which ones, if any, also resonate with you.

Obstacle #1: "I'm not sure what evangelism is."

PROBLEM: *Evangelism* is a word that is not normally used in any other context outside of Christianity. It's not something we would normally hear about at work, at home, or in conversations with friends outside of a Christian context. Let's start by clarifying what evangelism is and is not.

Simply put, evangelism is sharing the gospel of Jesus and our faith in him with someone who does not yet believe and therefore is not saved. We do this with the hope and intent that it would lead to that person having faith in Jesus. There are several points I (John) want to make about this.

First, sometimes evangelism is aimed at an unreached people group (which we have in America and online, by the way). In this case, the people have never heard of Jesus. In other cases, evangelism can involve someone who has grown up hearing about Jesus but is still not a believer. Evangelism is sharing the gospel with people who are not saved, whether they have heard of Jesus or not. There are many who "grew up in church" but still do not know Jesus and need to be evangelized. When you talk about Jesus, you may hear things like "I grew up in church," "My dad was a pastor," or even "I used to lead youth ministry." But you can't assume that they believe the gospel.

They may know the Church but not Jesus. They may assert confidence in the religious activities that they think matter to God but still fall short of authentic faith in Christ. Just because they can repeat the gospel does not mean they have repented and believed it. So sharing the gospel with them is still evangelism. Familiarity with the gospel does not mean faith in it.

Second, evangelism is characterized by the act of sharing the gospel, not the length of time spent doing so. Evangelism is the single conversation at a coffee shop, on a flight, or at a bookstore. Evangelism is also the many conversations with a person through years of an ongoing relationship.

> Familiarity with the gospel does not mean faith in it.

Third, since evangelism is the act of sharing the gospel, it is not the activity of reaping the harvest. It is not defined by whether or not someone gets saved. Even if the person does not give their life to Jesus after you shared the gospel, the act is still evangelism. Salvation is the *goal* of evangelism but not what defines it. Evangelism is sowing the seeds without being responsible for the quality or response of the soil.

Finally, evangelism is not inviting people to church. Earlier this year, I saw a Facebook ad about "how to teach your church to evangelize." But the goal of the organization was not to teach people to talk about Jesus; it was to teach them to talk about their church! Equipping people for evangelism is teaching them how to share their faith in Jesus Christ and how to share the gospel. If we are simply talking about our church or inviting people to it, then real work of evangelism is not being accomplished.

SUGGESTION: Let's get back to the biblical understanding and practice of evangelism. Only then will we see biblical results. I love the biblical examples of the focus and passion the apostles and the early Church had in sharing Jesus. Paul says, "But I do not account my life of any value nor as precious to myself, if only I may finish my course and the ministry that I received from the Lord Jesus, to testify to the gospel of the grace of God" (Acts 20:24, ESV). In 1 Corinthians 2:2 he writes, "For I decided to know nothing among you except Jesus Christ and him crucified" (ESV). After Philip baptized the Ethiopian eunuch, the Holy Spirit transported him to another city, and the first thing he wanted to do was preach Jesus! Acts 8:5 says, "Philip went down to the city of Samaria and proclaimed to them the Christ" (ESV). Let's just share Jesus and tell people that they can be reconciled to God through him!

> Let's get back to the biblical understanding and practice of evangelism.

Obstacle #2: "I don't know what to say or do!"

PROBLEM: No one likes to feel incompetent or inadequate. It's natural to avoid situations where we fear this happening. Although we agree that evangelism is important, we avoid opportunities that expose our inadequacies. Intentional avoidance can actually lead us to the point where we are so used to ignoring opportunities that we don't see them at all anymore—even when they are right in front of us.

Even as a pastor there were many times I was nervous about sharing Jesus because I worried that someone would ask a question I couldn't answer. We as Christians have heard countless teachings about why to share the gospel and what happens if we don't, but we have not heard much about how to do it. This lack of preparation prevents us from having conversations about Jesus and engaging our family and friends.

There are conversations and relationships that God has ordained for you to be in. These are opportunities he has given to *you*—not your pastor or church—to nurture and develop in a way that glorifies him. Without knowing how to respond to these "God moments," you may outsource *your* God moments to a paid clergy at your church or just ignore them and pass them up altogether. Part of being equipped as a digital missionary is confidently leaning into your personal responsibility to embrace those God moments yourself. God has plans and purposes for the relationships he has established for us. You can do more than just be a pipeline to the organized church. (Every believer should be connected to a local church, but I'm specifically speaking of jumping to that step prematurely because it's easier to let the church handle the responsibility of talking about Jesus.)

> Confidently lean into your personal responsibility to embrace those God moments yourself.

SUGGESTION: Let me (John) offer a couple suggestions, and you can decide how to apply them.

First, let's change our attitude about our ability to evangelize. If you or I were interviewing for a job and the interviewer asked us whether we knew how to do a certain task that we did not know how to do, the first thing most of us would say is, "I don't currently know how to do that, but I'm a fast learner!" Without knowing what would be involved in learning it, since we want the job, we would confidently communicate our ability to do whatever it took to learn how to do that task. You know, fake it 'til you make it! That's what I'm suggesting as an attitude toward evangelism. In this day and age, you can learn anything you want.

Second, just practice. Practice doesn't make perfect, but it does make progress and build confidence. Then you'll be able to "always be prepared to make a defense to anyone who asks you for a reason for the hope that is in you; yet do it with gentleness and respect" (1 Peter 3:15, ESV).

Obstacle #3: "What if evangelism is not my gift?"

PROBLEM: Although the word *evangelist* is mentioned several times in the New Testament, there is no clear definition given in Scripture. The role is also not clearly defined in many churches. In the book of Acts, Philip was called an evangelist (Acts 21:8), the apostle Paul told Timothy to do the work of an evangelist (2 Timothy 4:5), and Paul also listed evangelism as one of the five leadership gifts that Jesus has given the Church (Ephesians 4:11).

When many Christians think of evangelism, they think of Reverend Billy Graham or another well-known speaker. So it's really easy for the average Christian to quickly cancel out personal evangelism because "I am not like that" or even "there's no way God has called me to do that."

The truth of the matter is that there is the function, office, and gift of an evangelist in the church. This person has an above average ability to lead people to Christ. Jesus' plan for evangelists and other church leadership roles is made clear in Ephesians 4:11-12 (ESV): "And he gave the apostles, the prophets, the evangelists, the shepherds and teachers, to equip the saints for the work of ministry, for building up the body of Christ." Instead of leaders doing the evangelism themselves, another purpose of this gift is to equip the body of Christ to do evangelism.

SUGGESTION: It's true that there is a special gift of evangelism. But evangelism is something that every Christian should take part in (even though not every Christian will be an evangelist). Evangelism, the sharing of the gospel, is a privilege and responsibility for every Christian. It is something that every Christian is called and empowered by the Spirit of God to do. To those gifted in evangelism, your role is to help the body of Christ get better at it. Each believer is called to be a witness for Jesus (Acts 1:8), and each believer is called to share the message of reconciliation (2 Corinthians 5:18).

Obstacle #4: "I'm afraid I'll mess it up."

PROBLEM: It is really easy to make evangelism about us. One way we do so is by being too concerned about how well we share the gospel. For some, it can be performance-oriented. We want to see someone come to faith in Christ, but we put too much pressure on ourselves to make it happen. It's as if we believe that a person's salvation weighs completely on our flawless presentation of the gospel and our ability to quote every verse and answer every question that is asked. If we make evangelism all about our performance, then we are susceptible to performance anxiety, which can cause us to avoid opportunities to share Jesus. There were plenty of times when I felt a clear open door in a conversation but was unsure about my ability to clearly explain the gospel—so I didn't say anything.

SUGGESTION: Instead of focusing on our ability to present the gospel perfectly, we should focus on the Holy Spirit's ability to work perfectly in the heart of the hearer. Our role is to tell the good news. The Spirit's role is to convict of sin (John 16:8) and save souls.

Trust the Holy Spirit to do what you cannot. Trust him to fill in the gaps. I (John) remember hearing about a small group of believers evangelizing in a mall. They were led by "Sue" and were talking with a woman we will call "Jane." Sue was unsure of herself, and it showed in multiple ways. She stumbled over her words. She quoted Scriptures wrong. These mistakes made Sue even more insecure about sharing Jesus. Jane listened patiently and gave awkward smiles from time to time (which the group of Christians felt was Jane's way of being nice). As unsure of herself as she was, Sue committed herself to the conversation and shared core elements of the gospel. Jane ended up accepting Christ! But here's why I am sharing this story: Jane said, "I felt loved while you were speaking. That's how I knew that what you were saying was real." Now you and I know that the love she felt was God moving in her heart. That's my point! Listen, God chooses to work with flawed, imperfect people. So he knows he has to fill in some gaps for us. Don't let performance anxiety prevent you from sharing the gospel. You can do it! Instead of asking, "What if I mess it up?" ask yourself, "What if God helps me?" You can trust him to be with you! We are naturally more comfortable in areas where we feel competent, but our obedience will always produce more Kingdom results than our competence.

Obstacle #5: "I'm afraid of rejection."

PROBLEM: Another way our self-focus shows up is in our fear of rejection. When we are more concerned about how people view us, what they think of us, or whether they will accept us or reject us, those worries can certainly discourage us from sharing the gospel. We can't be afraid we will ruin a relationship by shifting a conversation to spiritual things.

Sometimes Christians get accused of being religious—acting better than others by telling others how to live, being judgmental, or communicating that "our way is the only way." There are so many negative stereotypes of Christians that the possibility of being categorized in one of those stereotypes can dissuade us. Not everyone is open to the truth of the gospel regardless of how it is presented. Sometimes it's a timing issue where you plant a seed that God will water later. Sometimes you share with a person whose heart is not and will never be open. Other times you will share the gospel and witness the joy of someone giving their life to Christ.

SUGGESTION: Sharing the gospel does risk rejection for many more reasons, and the fear can be very real. Regardless of the reasons, I want to offer some suggestions. Acts 4:18-20 (ESV) tells us, "So they called them and charged them not to speak or teach at all in the name of Jesus. But Peter and John answered, 'Whether it is right in the sight of God to listen to you rather than to God, you must judge, for we cannot but speak of what we have seen and heard.'"

Later in that chapter, the apostles prayed for boldness in light of threats they were receiving while preaching the gospel (Acts 4:29), but they still preached it. Sometimes our self-focus shows up when we are more concerned about our experience of sharing the gospel (our fear and anxiety) instead of someone else's experience of hearing it. Our tendency toward self-preservation can sabotage our efforts. What I admire about the apostles is that instead of shrinking back from rejection, instead of praying it would go away, they prayed for boldness to not let it stop them. Their prayer for boldness lets us know that at some level they were impacted by the threats and the rejection. They prayed for supernatural emotional support to strengthen them—and they received it.

> Our obedience will always produce more Kingdom results than our competence.

Obstacle #6: "I just don't have the desire to share Jesus."

PROBLEM: When I thought about this obstacle, I (John) considered how many other things people willingly, confidently, and joyfully share (even with strangers) without needing encouragement to do it.

We take pictures of our food and post them online. We share our excitement about a movie we want our friends to experience. I have never seen an usher stand at the exit door of a theater and say, "If you enjoyed the movie, please tell your friends." They don't have to do that. If I enjoyed the movie, I will naturally tell everyone. We gladly share joyful experiences because we want others to experience what we have experienced.

Why wouldn't we want to talk about Jesus that way? If we are singing songs to him, committing our lives to him, worshiping him, and more, why wouldn't we want to share Jesus with people? What's the hindrance? You can probably see where I'm going.

Perhaps one reason you don't share the gospel is that you haven't been changed by it. Maybe you haven't experienced the transformational power of the gospel. In his book *Transforming Fellowship: 19 Brain Skills That Build Joyful Community*, Chris Coursey said:

> The result of all healing and maturity is evangelism. Having once been transformed to a full and joyful life, it is beyond imagination to consider keeping this good news from others. Evangelism is nothing other than spreading the good news of what has so transformed our lives for the better. Telling people what we believe is of relatively little value or power compared to the radiance of profoundly transformed lives. Most people do not evangelize because they have no good news to share.[27]

I absolutely agree. Now if this resonates with you, don't panic. Let me explain how this can happen and where to go from here.

> Perhaps one reason you don't share the gospel is that you haven't been changed by it.

Throughout my ministry, I've seen that many Christians view believing the gospel as merely what *starts* their life with God. They look for something else to help them learn to grow in their relationship with God. This hinders spiritual growth because there *is* nothing else. In the words of J.D. Greear, "The gospel, however, is not just the diving board off of which we jump into the pool of Christianity; it is the pool itself. It is not only the way we begin in Christ; it is the way we grow in Christ. As Tim Keller says, the gospel is not just the ABCs of Christianity, but it is the A–Z."[28]

Maybe you are still on the diving board and have not jumped into the pool. Greear said the gospel is not just the message that gets you saved. It's the message that helps you live. We never outgrow our need to understand the gospel more. There are so many Christians who are frustrated with their life with God, frustrated with their level of spirituality, yet they are expected to share the joy of the gospel when many of them would love to first experience that themselves.

Many Christians are not "satisfied customers" of the gospel and have not experienced the joy of Christ that would naturally make them want to share it. It is assumed that you are experiencing the richness of a relationship with God—but maybe you're not. Giving someone tools to do something they do not have the desire to do misses the mark. Only when there is a desire to do something do tools have value. That's when equipping can truly happen. The fact that you are reading this book means you desire to share the gospel with others, specifically online. Your desire might not be as strong as you would like, but it was strong enough for you to pick up this book. And we're glad you did!

SUGGESTION: The solution is discipleship. Get discipled. Find someone you can meet with who will help you dive into the pool of the gospel! We often give a pamphlet with the Gospel of John to new believers, but perhaps you should check one out yourself. Go to your church and ask for their new believer material. It usually breaks down the fundamentals of our faith. Just start there. Perhaps you have been in church for 20 years. Are you comfortable explaining the core elements of the gospel? If not, take the humble approach of acting like a new believer, and explore everything you can about the gospel. As you do, you will discover the joy, peace, security, and transforming power of the gospel of Jesus. Don't try to force yourself to do evangelism. Go and get the joy of the Lord, and let it propel you into evangelism. As you grow in the gospel, you will discover the truth of what Bill Thune said in *The Gospel-Centered Life*: "When the gospel is at work in you, it will also be at work through you."

I tried this approach in Hearts Like His. For months, all my wife and I did was teach the gospel. As a result, one person started an online Bible study with family members and has recently started a house church. Another person said, "I've been saved for over 40 years and have shared my faith maybe once. Since being in Hearts Like His, I have shared my faith 10-15 times this year!" He even called his hometown pastor and said, "I'm really forgiven in Christ. I didn't have to keep mowing the church's lawn when I messed up!" He shifted from feeling like he owed good deeds for his bad deeds to knowing that the gospel provides

> Go and get the joy of the Lord, and let it propel you into evangelism.

complete forgiveness, and when the truth set him free it resulted in a natural outward expression of evangelism.

Obstacle #7: "I preach the gospel without words."

"Preach the gospel at all times. When necessary, use words" (Assisi).[29] I (John) have heard this quote for years. "It is always attributed to St. Francis of Assisi—founder of the Franciscan Order—and is intended to say that proclaiming the gospel by example is more virtuous than actually proclaiming with voice."[30] However, there are several problems with this quote. First, Francis of Assisi never said that. According to Mark Galli, senior managing editor at *Christianity Today* and biographer of Francis, "No biography written within the first 200 years of his death contains the saying. It's not likely that a pithy quote like this would have been missed by his earliest disciples. Second, in his day, Francis was known as much for his preaching as for his lifestyle."[31] Not only did he not say this, but he did not live it. I was shocked to discover that he was:

> Sometimes preaching in up to five villages a day, often outdoors. In the country, Francis often spoke from a bale of straw or a granary doorway. In town, he would climb on a box or up steps in a public building. He preached to serfs and their families as well as to the landholders, to merchants, women, clerks, and priests—any who gathered to hear the strange but fiery little preacher from Assisi.[32]

So St. Francis didn't say it *or* live it, but many Christians use it as a reason not to share Jesus. If this belief hinders your in-person evangelism, it would definitely hinder your online approach as well. Let's unpack this for a moment or two.

Without question, our lives should testify to the truth and the power of the gospel we preach. One of many verses that makes that clear is James 2:18 (ESV): "But someone will say, 'You have faith and I have works. Show me your faith apart from your works, and I will show you my faith by my works.'" Another Scripture is 1 Peter 3:1-2 (ESV):

"Likewise, wives, be subject to your own husbands, so that even if some do not obey the word, they may be won without a word by the conduct of their wives, when they see your respectful and pure conduct."

But there are two problems with interpreting these verses in a way that supports "example only" evangelism. First, Peter was not describing the best way for everyone to evangelize. In Peter's day, women did not have the voice or authority to teach or lead their husbands at all—especially spiritually. Peter said that although wives didn't have that kind of influence, they could at least live out what they believed and still have an impact. Second, if you continue to verse 15, Peter says clearly to always be "prepared to make a defense to anyone who asks for a reason for the hope that is in you." In other words, when sharing the gospel of Jesus words *are* necessary!

Too many Christians use this "my life will tell it" approach so much that they are actually unprepared to share the gospel, feel no need to learn, and are not even looking for these opportunities. This promotes evangelistic passivity. There is nothing passive about preaching and sharing the gospel. This attitude contributes to the crisis on hand regarding our lack of evangelism.

I recently consulted with a wonderful and caring group of church leaders about their community outreach, lack of conversions, and decline in numbers. Unfortunately, this faulty Assisi quote was their church motto. They felt all they had to do was be nice and charitable to people, and God would do the rest. None of them felt the need to actually ever tell people about Jesus. I helped them see the connection in their lack of people coming to Christ. No one walks away from receiving soup, coats, or a backpack saying, "Now I get it! Jesus is the Son of God who died on the cross for my sins, and his blood is the only way I can be forgiven and be in right standing with God." Soup, backpacks, and showing love can bless the community, relieve suffering, provoke questions, and inspire curiosity about our faith—but they do not preach it.

SUGGESTION: Our transformed lives and our works will show the reality and power of the gospel message, but they do not explain the gospel message. Our lives point to the gospel; they don't preach it. Preaching and sharing the gospel always requires words. Sharing the gospel and living the gospel go together—but they are not the same. Yes, our lifestyle can and should provoke questions, but the answer to those questions lies in sharing the gospel.

No one lived a life better than Jesus, and even *he* still had to preach: "'The Spirit of the Lord is upon me, because he has anointed me to proclaim good news to the poor. He has sent me to proclaim liberty to the captives and recovery of sight to the blind, to set at liberty those who are oppressed, to proclaim the year of the Lord's favor'" (Luke 4:18-19, ESV). Most of what Jesus was anointed to do involved preaching and proclaiming.

Likewise, he told his disciples, "And proclaim as you go, saying, 'The kingdom of heaven is at hand.' Heal the sick, raise the dead, cleanse lepers, cast out demons. You received without paying; give without pay" (Matthew 10:7-8, ESV). Healing the sick, raising the dead, and casting out demons were powerful miracles that served people, but notice how these were intended to accompany the message of God's Kingdom, not replace it. As Paul, who frequently and passionately taught about the importance of godly living, said, "How then will they call on him in whom they have not believed? And how are they to believe in him of whom they have never heard? And how are they to hear without someone preaching?" (Romans 10:14, ESV).

Even though Francis of Assisi never said it, let me change the quote to "Preach the gospel because it's always necessary to use words, especially online."

As we look at perceived obstacles that hinder evangelism, there are also perceived misconceptions of digital and metaverse communities. Here are just a few.

Metaverse Misconception #1: Everyone is fake online.

> No one walks away from receiving soup, coats, or a backpack saying, "Now I get who Jesus is!"

PERCEPTION: This concern isn't wrong, although "everyone" may be an exaggeration. There is a perception that people are different digitally than physically or in person. Platforms like Instagram highlight the best of people's lives. Rarely do people use these platforms and tools to highlight the dark or lonely parts. The perception is that "I don't really know even the people I 'know' digitally because I've never met them physically (or it's been a long time since I've done so)."

REALITY: If we're being fair, we could just as easily say that everyone is fake physically. In fact, there's a growing percentage of people who have walked away from physical church buildings because of this perceived fakeness. (Many of these de-churched people are reconnecting to God and the Church digitally.) While the perception is that you don't really "know" people digitally, many people are open to genuine, honest relationships digitally and in the metaverse.

CHALLENGE: The opportunity for the digital missionary is to not question the relationships we're developing digitally or virtually. Instead, we should assume the best and model what an authentic relationship could look like. The more real you are with someone digitally or in the metaverse, the more real the other person will be in return.

Metaverse Misconception #2: People have to be in the same room to have relationships or community.

PERCEPTION: The internet is not relational. All people do is yell at each other online. To have true relationships, you have to be together in physical space. Ministry happens best physically. People do not want to talk about Jesus digitally. I'm not really comfortable talking about Jesus at all, much less digitally.

REALITY: People can be quite relational digitally, and these relationships can point to Jesus. Digital is a great environment in which to share Jesus, and people are open to conversations about Jesus in digital spaces and in the metaverse.

CHALLENGE: There is an ongoing conversation about whether physical or digital environments are better for discipleship. For what it's worth, I (Jeff) think this physical/digital conversation is a distraction at this point. The fact is that we have access both digitally and physically to people who need Jesus. We should be taking every opportunity we can to connect people to Christ's saving grace, so let's consider both avenues! Whether digital evangelism is better or worse for discipleship, it is *as valid* and *as necessary* as physical evangelism.

Metaverse Misconception #3: Nothing of God can come from virtual reality.

PERCEPTION: Virtual reality is a dark place. Satan has a strong grasp on virtual reality. God cannot, nor should he, have anything to do with virtual reality.

REALITY: It's true. Satan does have a grasp on virtual reality. However, tools like virtual reality and social media are not "of Satan" any more than they are "of Jesus." In fact, we often describe these tools as being "agnostic." A hammer is not Satanic, but it can be utilized in evil or in godly ways.

CHALLENGE: The current darkness of virtual reality is not taking place because God cannot or will not exist in virtual space. The current darkness is taking place because we have not taken God into virtual reality. Even in virtual space, God can use us as missionaries bringing light into the darkness.

Metaverse Misconception #4: You don't know who people really *are* in the metaverse!

PERCEPTION: Virtual reality is a make-believe world where people pretend to be something they're not. It's a gamers' playground with no merit or grounding in reality.

REALITY: As I (Jeff) discussed in my recent book *VR & the Metaverse Church: How God Is Moving in This Virtual, Yet Quite Real, Reality*, yes . . . the metaverse is a virtual world, but the people interacting inside it are quite real. And while we often don't know what their actual names are or what they look like, virtual reality is filled with people who want to talk. In fact, some of the best conversations I've had about Jesus have been with avatars. The anonymity of virtual reality is actually a strength that allows people to talk openly without fear of being judged.

CHALLENGE: This mindset is somewhat different from physical ministry, where we want people to fill out that connection card as quickly as possible. In the metaverse, as well as digitally, sometimes it makes more sense to build a relationship without knowing that level of detail. Truthfully, there's a lot of relationship potential even when you don't know the name listed on someone's birth certificate.

> The anonymity of virtual reality allows people to talk openly without fear of being judged.

Metaverse Misconception #5: God cannot enter the metaverse!

PERCEPTION: God cannot interact in or even enter virtual reality. Virtual reality cannot be of God because VR is inherently man-made.

REALITY: Ask Christians who are active in virtual reality spaces. We see God moving there regularly. The idea that God cannot interact with people in VR is ridiculous. Do people think this way because God doesn't own a VR headset? (Sarcasm implied and hopefully understood.) Biblically, of course, there's nothing to validate or refute this idea because virtual reality is not mentioned in the Bible (obviously). That being said, let's remember that since the time of Adam and Eve, man has unsuccessfully tried to hide from God.

"'Can a man hide himself in secret places so that I cannot see him?' declares the Lord. 'Do I not fill the heavens and the earth?' declares the Lord" (Jeremiah 23:24, ESV). If God is powerful enough to oversee the heavens and the earth, I'm pretty sure he can rule over VRChat.

If virtual reality is in fact the technology that allows man to hide from God, then we've got bigger problems theologically than God not hearing our prayers.

CHALLENGE: Recognize that the Holy Spirit is fully capable of moving in these physical, digital, and metaverse spaces. Pray about following God's lead as you discern who to B.L.E.S.S. and how to best do so.

> If God is powerful enough to oversee the heavens and the earth, I'm pretty sure he can rule over VRChat.

Remember, spiritual warfare is real. People are blinded by our enemy. Paul said in 2 Corinthians 4:4 (ESV), "In their case the god of this age has blinded the minds of unbelievers, to keep them from seeing the light of the gospel of the glory of Christ, who is the image of God." Certainly, our enemy would love nothing more than for you to believe that sharing the gospel is not for you.

• • •

Now that we have these obstacles and misconceptions out of the way, let's move forward. My prayer is that if you were wondering, "Should I do this?" your question has now changed to, "What's the best way that I can do this?" The B.L.E.S.S. framework is a perfect tool. So let's dive in!

CHAPTER 3

To B.L.E.S.S., Always Begin in Prayer

DAVE SAYS

Crazy things begin with prayer. Holocaust survivor Corrie ten Boom once said, “We never know how God will answer our prayers, but we can expect that he will get us involved in his plan for the answer.” I believe that! Sometimes when God gets us involved, outlandish and powerful stuff begins to happen.[33]

Successful Evangelism Begins with Prayer

CHARLES FINNEY IS CONSIDERED BY MANY to be the greatest revivalist of the Second Great Awakening. He is called the “father of modern revivalism” by some historians, and he paved the way for later mass evangelists like Dwight L. Moody, Billy Sunday, and Billy Graham.[34] Many people have heard of Finney because of the thousands who responded to the gospel through his preaching. However, very few have heard of an Episcopal priest named Father Daniel Nash from New York. Finney and Nash had quite a partnership in ministry. Not only did Father Nash travel with Finney, but often he would go before him.

“When God would direct where a meeting was to be held, Father Nash would slip quietly into town and seek to get two or three people to enter into a covenant of prayer with him. Together they would begin to fervently pray for God to move in the community."[35] They would be in the town two to three weeks ahead of Finney and pray. Then they would let Finney know when the town was ready. As powerful as his evangelism ministry was, Finney did not go into a town until he began with prayer. "Finney stated, 'I did the preaching altogether, and brother Nash gave himself up almost continually to prayer.'"[36] Nash's tombstone reads, "Laborer with Finney, Mighty in Prayer." After Nash died in 1831, Finney ended his traveling ministry and began to pastor a church. He was no longer getting the same results.

I (John) love the confident partnership Father Nash had with the Holy Spirit. It shows his understanding of the role of prayer in producing Kingdom results, his obedient submission to the call to prayer, and his passion for lost souls. Evangelism must always begin with prayer.

Prayer removes ego from the situation, placing you in a posture of humility.

What if we prayed for souls, sermons, direct messages, and chat conversations the way Daniel Nash prayed for the people in those cities and towns? What if we started praying for the hearts of unsaved people we meet online to be open to receiving the good news of Jesus? For something to shift in the lives of those we hope to B.L.E.S.S., something first must shift in us.

Prayer Is the Same in Any Environment

Beginning with prayer is essential—in any environment! Physical, digital, social media, virtual reality . . . in any of these areas, starting out with a prayerful focus is important because:

1. It invites the Holy Spirit to lead the conversation.
2. It takes you out of the driver's seat.

Prayer removes ego from the situation, placing you in a posture of humility.

Communicating with the creator of the world forces us into a place where we must acknowledge that God is greater than us. And that is right where we need to be. Our goal as digital missionaries is not to orate some incredibly convincing presentation. It is to assist the Holy Spirit in connecting with people. The success of evangelism is not based on our abilities but on the Holy Spirit—so why would we not start our conversations humbly in prayer?

Jesus Began with Prayer

Beginning with prayer is something Jesus modeled for us. He began his ministry with 40 days of fasting (Matthew 4, Luke 4). The night before he chose his disciples, he spent the night in prayer (Luke 6:12). Jesus often began his day in prayer, "And rising very early in the morning, while it was still dark, he departed and went out to a desolate place, and there he prayed" (Mark 1:35, ESV). If Jesus, who *is* God, thought it important to begin with prayer, how much more in need of this are we?

Beginning with Prayer Prepares You

Anyone who has been praying for long will tell you that when you begin with prayer, one of the first things prayer will change is you. Let's look at some things you can pray about for yourself and then for your specific missional focus.

Pray to Grow in the Gospel

As you prepare to share the good news that God forgives sin through faith in Christ, pray for a deeper personal understanding and experience of the transforming power of the gospel. The gospel is the basis for our growing love of God and our conformity to the image of Christ. Grow in your understanding of his complete forgiveness of your sins through your faith in Jesus. Grow deeper in your understanding of his grace that sets us free from religiously striving to earn God's acceptance and approval. Learn to stand firm in the freedom for which Christ has set you free (Galatians 5:1)!

Our prayer is "that you, being rooted and grounded in love, may have the strength to comprehend with all the saints what is the breadth and length and height and depth, and to know the love of Christ that surpasses knowledge, that you may be filled with all the fullness of God" (Ephesians 3:17-19, ESV). As you grow in your understanding and experience of God's love as demonstrated in the gospel of Jesus, may it ignite a fresh fire and passion for sharing it.

Pray for Boldness

Earlier, I (John) mentioned how Peter prayed for boldness in the face of threats (Acts 4:29). As Paul was asking the Ephesian church to pray for all the saints, he added: "and also for me, that words may be given to me in opening my mouth boldly to proclaim the mystery of the gospel, for which I am an ambassador in chains, that I may declare it boldly, as I ought to speak" (Ephesians 6:19-20, ESV).

If the apostles who met Jesus in the flesh needed to pray for boldness, then it should definitely be on our list. In America, we have not seen the kind of persecution early Christians experienced in the book of Acts, but you may have a different reason to pray for boldness. Perhaps your challenge was mentioned in our list of obstacles and misconceptions, or maybe it's something closely related. Whatever the reason, pray for boldness as you move through each step in the B.L.E.S.S. framework.

Pray to Have a Heart for the Harvest

Matthew 9:35-38 (ESV) records Jesus' prayer request:

> And Jesus went throughout all the cities and villages, teaching in their synagogues and proclaiming the gospel of the kingdom and healing every disease and every affliction. When he saw the crowds, he had compassion for them, because they were harassed and helpless, like sheep without a shepherd. Then he said to his disciples, "The harvest is plentiful, but the laborers are few; therefore pray earnestly to the Lord of the harvest to send out laborers into his harvest."

You cannot "pray earnestly to the Lord of the harvest to send out laborers" without considering yourself to be one of them. Jesus knew that one day he would pass the baton to his disciples. They needed to have a heart for the harvest, his heart for the harvest. Jesus literally told his disciples to pray for *you*! This prayer request came from his compassion for the crowd because of their helplessness. Jesus did not intend for God to send just anyone, but to send laborers who saw what he saw and felt what he felt about this harvest.

Without compassion for those who are not saved, we are prone to be judgmental, overly critical, and condemning, which leads us to use religious, manipulative scare tactics. Without compassion, sharing the gospel becomes either a religious work that you do out of a sense of obligation or something that you avoid altogether. However, with a compassionate heart for the harvest (whether in physical, digital, or metaverse space), your mission becomes an ongoing joy for you even when it's hard. With a heart for the harvest (especially the digital harvest), you will have a spiritual readiness and sensitivity when God leads you to seek the lost or sends the lost to you. Let's talk more about what that looks like.

> You cannot "pray earnestly to the Lord of the harvest to send out laborers" without considering yourself to be one of them.

Seeking the Lost, Welcoming the Lost

Jesus said he came "to seek and to save the lost" (Luke 19:10, ESV)—and he still has that same mission and desire. But now, he does it through us! "As the Father has sent me, even so I am sending you" (John 20:21, ESV).

Some of Jesus' encounters with people occurred because he intentionally sought them out. As the seeking Savior, he went out of his way to meet people. Although Jews typically went around Samaria, he went through Samaria to meet a woman at the well. He sailed across the Sea of Galilee to set a man free of demonic possession. He invited himself to a meal at Zacchaeus' house so he could spend time with him.

Other encounters were with people who approached him. As the welcoming Savior, Jesus helped people who came to him—for example, the four men who lowered their friend through the roof to be healed, the centurion whose servant was sick, Jairus' dying daughter, the woman with the issue of blood, the Syro-Phoenician woman whose daughter had a demon, blind Bartimaeus who cried out for mercy . . .

and the list goes on. He welcomed women. He welcomed other ethnicities. He welcomed social outcasts. He welcomed those who were up and those who were down.

A heart for the harvest means being ready to seek and to welcome the lost. What does this posture look like? I'm glad you asked.

Asking a Different Question

I (John) believe that having a heart for the digital harvest begins with asking a different question than the one we typically ask ourselves. It's a question made clear in an excerpt from Dr. Martin Luther King, Jr.'s "Mountaintop Speech" on April 3, 1968, the evening before he was assassinated. As Dr. King shared about the parable of the Good Samaritan, he explained some possibilities as to why the priest and the Levite passed by the man on the side of the road:

> I'm going to tell you what my imagination tells me. It's possible that those men were afraid. You see, the Jericho Road is a dangerous road. I remember when Mrs. King and I were first in Jerusalem. We rented a car and drove from Jerusalem down to Jericho. And as soon as we got on that road I said to my wife, "I see why Jesus used this as the setting for his parable." It's a winding, meandering road. It's really conducive for ambushing.
>
> You start out in Jerusalem, which is about 1200 feet above sea level, and by the time you get down to Jericho, 15 or 20 minutes later, you are 2200 feet below sea level. That's a dangerous road! In the days of Jesus, it came to be known as The Bloody Pass, so it's possible that the priest and the Levite looked over at that man on the ground and wondered if the robbers were still around. It's possible that they felt that the man on the ground was merely faking. And he was acting like he had been robbed and hurt in order to seize them over there, to lure them there for a quick and easy seizure. So the first question that the priest asked, the first question that the Levite asked, was, "If I stop to help this man, what will happen to

> me?" But then the good Samaritan came by, and he reversed the question: "If I do not stop to help this man, what will happen to him?" That's the question before you tonight!

Having a heart for the digital harvest means asking a different question. Instead of asking, "What will happen to me if I share Jesus?," we should be more concerned with "What will happen to them if I don't?"

Pray About Who and Where

I (John) have been guilty of confidently starting ministry endeavors without prayer. I thought that any ministry idea was automatically approved by God. Because I was doing it for him, I felt like I did not need to pray over everything. Right? Oh, how I know better now! Although we may have a general idea of his plan, there are specific things that definitely require prayer. God wants us to do this work together with him. He wants to guide us. We are not just "workers" *for* God, we are fellow workers *with* him (1 Corinthians 3:9), and he leads the way.

One thing that characterizes a missionary mindset is the choice of a missional target. Missionaries intentionally choose a group of people to build relationships with and plant the gospel. When you pray for the harvest, who is your heart crying out for? What group of people do you have a soft spot for? Who do you feel drawn to help? Is it single parents, new believers, young adults, empty nesters, gamers, pastry chefs, fishermen, business owners, fitness fanatics, outdoor sports enthusiasts, married couples, people who are grieving, parents, educators, foster youth, the recovery community . . . I could go on, but you get the point. These are people you can intentionally "seek" digitally. Although we are called to love everyone, we are not led to focus on everyone.

When the Spirit highlights a particular group or person, remember that God has already been working in that group or in that person's life. We are joining in what he is already doing. Beginning with prayer reminds us of this truth.

Once you are clear on the who, the where becomes clear. Determine where you will find this group or person in digital and metaverse spaces, and go there! Different demographics are better suited for different platforms. For example, younger generations use Instagram and TikTok more than Facebook. Gamers are definitely in virtual reality spaces. The people you are called to (not your comfort zone) determine the places you go. That's the missionary mindset. There may even be a more specific focus or group within a platform's culture. For example, "as of August 2022, over 10 million groups were on Facebook, with over 1.8 billion people using them each month."[37] Facebook groups are a great opportunity for digital missionaries (like you!) to build relationships and start conversations among people who already have shared interests.

The who, the where, the when, the what, the how, the why—it is all important, and it all requires prayer. Never rely on your own plan so much that you forget whose work you are doing.

Some Fish Are Ready to Jump in the Boat

One of the main ideas I (John) want you to take away from this book is knowing that evangelism is our Father's business. He is eager to help you be successful. Know beyond the shadow of a doubt that God does not wish "that any should perish, but that all should reach repentance" (2 Peter 3:9, ESV). He *wants* to save people. I'm not saying that people will be saved in every one of our conversations, but we can certainly see more fruit when we allow him to lead us.

Jesus told his disciples on two occasions where to put their nets so they could catch fish (Luke 5:1-11, John 21:6). He can direct you in the same way. I want to highlight two biblical examples where the Spirit of God led believers to people who were ready to hear the gospel and respond in faith. They were fish who were ready to jump in the boat. The Spirit is just as eager to lead and help you. When you follow the Spirit's leading, he is bringing you to a place where he is already at work. As you read these biblical accounts, pay attention to three themes: the Spirit's involvement, the believer's obedient partnership, and the recipient's readiness and response. John 6:44 (ESV) says that "no one can come to me unless the Father who sent me draws him. And I will raise him up on the last day." Pay attention to how God sends and God draws in these two accounts.

Philip and the Ethiopian

One of the most notable examples of someone who was already ready is the story of Philip and the Ethiopian eunuch in Acts 8:26-29, 34-40 (ESV):

> Now an angel of the Lord said to Philip, "Rise and go toward the south to the road that goes down from Jerusalem to Gaza." This is a desert place. And he rose and went. And there was an Ethiopian, a eunuch, a court official of Candace, queen of the Ethiopians, who was in charge of all her treasure. He had come to Jerusalem to worship and was returning, seated in his chariot, and he was reading the prophet Isaiah. And the Spirit said to Philip, "Go over and join this chariot."
>
> And the eunuch said to Philip, "About whom, I ask you, does the prophet say this, about himself or about someone else?" Then Philip opened his mouth, and beginning with this Scripture he told him the good news about Jesus. And as they were going along the road they came to some water, and the eunuch said, "See, here is water! What prevents me from being baptized?" And he commanded the chariot to stop, and they both went down into the water, Philip and the eunuch, and he baptized him.
>
> And when they came up out of the water, the Spirit of the Lord carried Philip away, and the eunuch saw him no more, and went on his way rejoicing. But Philip found himself at Azotus, and as he passed through he preached the gospel to all the towns until he came to Caesarea.

> When you follow the Spirit's leading, he is bringing you to a place where he is already at work.

amazon.com

SNtnK2Lj8R

Purchase Order #: AS2163495
Order ID 113-9430359-0224247 - Order of November 2, 2026

Qty.	Item
1	**Sharing Jesus Online: Helping Everyday Believers Become Digital and Metaverse Missionaries** Harris, Dr. John --- Paperback **1624241123** 1624241123 9781624241123

Return or replace your item
Visit Amazon.com/returns

0/NtnK2Lj8R/-1 of 1-//MKE6-CART-CO/next-1dc/0/1102-23:00/1102-18:38

SmartP

A gift for you

Enjoy your gift.

The eunuch was not interested in a debate, nor was he resistant in any way. He was reading the Word of God and wanted to know more. After Philip shared the gospel, the man was ready to be saved and baptized. He didn't even want to wait until baptism Sunday so he could invite his family and take pictures. He wanted to get baptized right away! He was ready. God made him ready.

Paul in Macedonia

Another example is when the apostle Paul attempted to share the gospel in different regions but God redirected him. Again, pay close attention to God's intervention and guidance for Paul and his crew in Acts 16:6-10 (ESV):

> And they went through the region of Phrygia and Galatia, having been forbidden by the Holy Spirit to speak the word in Asia. And when they had come up to Mysia, they attempted to go into Bithynia, but the Spirit of Jesus did not allow them. So, passing by Mysia, they went down to Troas. And a vision appeared to Paul in the night: a man of Macedonia was standing there, urging him and saying, "Come over to Macedonia and help us." And when Paul had seen the vision, immediately we sought to go on into Macedonia, concluding that God had called us to preach the gospel to them.

What was the result of Paul and his companions going to the region of Macedonia? Luke, the author, tells us in Acts 16:13-15 (ESV):

> And on the Sabbath day we went outside the gate to the riverside, where we supposed there was a place of prayer, and we sat down and spoke to the women who had come together. One who heard us was a woman named Lydia, from the city of Thyatira, a seller of purple goods, who was a worshiper of God. The Lord opened her heart to pay attention to what was said by Paul. And after she was baptized, and her household as well, she urged us, saying, "If you have

> judged me to be faithful to the Lord, come to my house and stay." And she prevailed upon us.

Did you catch what happened? "The Lord opened her heart to pay attention to what was said by Paul." This is what happens when we co-operate with the Holy Spirit. It was not Paul's responsibility to prepare Lydia's heart. It was the Lord's. The Lord was working on both sides. So why would he do that? Because he wants his children home. He's at work in the hearts of those who are searching. He's guiding us to them and them to us. I (John) really want you to get this part so you can relax! This is *his* show, and he is giving you a few lines to say on the stage of someone else's life. Just say your lines! He will handle the rest. He wants to include you because he loves you, but he is the one at work. I hope that takes some pressure off you. You get to enjoy this process! You get to watch the Father bring his kids home. Don't be anxious. Be led by his Spirit, and he will go before you!

There were specific people and places that God directed Philip and Paul to go. He will do the same thing as you share the gospel online. Stewart Freeman, the VRChat director at Cornerstone VR, shared this experience:

STORY

There was a week earlier this year when God used me to lead four people to Christ. One of these four [instances] was unusual. I was asleep. God woke me up at 3 a.m. In the pit of my stomach, it was like the Holy Spirit was telling me to get on VRChat right now. So I did. I scrolled through the worlds while constantly praying, "God, where do you want me to go?" Then I found it. There was a guy that had come to the church once. He was in a sleeping world. Many go to sleep in VRChat, for the comfort of the company of others and the ambiance many worlds bring. He used the public setting for the sleeping world he was in so friends could join, but he was alone. So I joined and started talking to him; he was very depressed. Setting the world to "public" in the middle of his depression was a cry for help. I had the opportunity to tell him the gospel, and he became saved.

In virtual reality, God is still working. He graces me to be able to see it every single day.

God can and will direct you. Not only that, but he is looking forward to it! If we are to see more success in evangelism and missions, we must be sensitive and submissive to the guidance of God's Spirit. Beginning with prayer positions our hearts to do just that.

"No, Not Them, Lord!"

Our society has divided people into different categories and labeled some as more valuable than others. These categories and values do not exist in the heart of God, nor should they exist in our hearts. Our challenge is to rise above those divisions (even ones taught to us by our families) and see people with God's heart. Our beliefs about race, politics, class, gender, and sexual orientation can show up when God sends us to a person or people group about whom we may have wrong or even harmful beliefs. What's crazy is that you might not even think you have that issue until God sends you and some weird resistance rises to the surface. Sometimes obeying God's call brings us to deeper levels of self-awareness. It's easy to avoid people in person, and *much* easier to avoid them online. We can just scroll past them without them even knowing we saw them. But what happens when God highlights someone to you online who you would much rather avoid? On your side of the screen there is no accountability, as there would be in an in-person interaction where body language and facial expressions are visible. The truth is that God does not respect our prejudices.

Jonah resisted God's cross-cultural ministry call to the people of Nineveh. God did not change the call—he changed Jonah. He gave Jonah time to think in a private three-day retreat in a fish's digestive system (Jonah 1:17). The result was that Jonah obeyed and preached to Nineveh, and Nineveh repented. When Peter resisted God's idea for him to go to the Gentiles (non-Jews, a different race of people), God did not change the plan. He changed Peter. He gave Peter three visions to prepare his heart for ministry to the Gentiles at Cornelius' house (Acts

10:9-16). The result was the Holy Spirit being poured out on the Gentiles as well as on the Jews on the day of Pentecost.

Do you see the pattern? Don't look just at how these men struggled with their assignments. Look at how God did not abandon them but instead journeyed with them to the point of clarity and a change of heart. These stories are not just about their stubborn resistance, but also God's patient persistence. Expect him to be patient and persistent with you too. He won't change the call. He will change you, giving you a heart for the harvest—like his own.

> Sometimes obeying God's call brings us to deeper levels of self-awareness.

We can clearly see God's heart in the ministry of Jesus. Jesus gave us a master class on crossing racial, economic, gender, even religious barriers in the encounter he initiated with the woman at the well. Is there a group of people to whom you would be challenged to show kindness? Instead of praying for God's mercy on them, would you prefer God's wrath as Jonah did for Nineveh?

Another example was Jesus' encounter with Zacchaeus. Jesus responded to Zacchaeus in a completely different way than how the crowd expected or wanted him to respond. Luke 19:1-10 (ESV) says Jesus was passing through Jericho:

> And behold, there was a man named Zacchaeus. He was a chief tax collector and was rich. And he was seeking to see who Jesus was, but on account of the crowd he could not, because he was small in stature. So he ran on ahead and climbed up into a sycamore tree to see him, for he was about to pass that way. And when Jesus came to the place, he looked up and said to him, "Zacchaeus, hurry and come down, for I must stay at your house today." So he hurried and came down and received him joyfully. And when they saw it, they all grumbled, "He has gone in to be the guest of a man who is a sinner." And Zacchaeus stood and said to the Lord, "Behold, Lord, the half of my goods I give to the

> poor. And if I have defrauded anyone of anything, I restore it fourfold." And Jesus said to him, "Today salvation has come to this house, since he also is a son of Abraham. For the Son of Man came to seek and to save the lost."

There are several observations worth mentioning here. First, Zacchaeus was hated by many in the crowd, but that did not shape Jesus' view of him. Tax collectors were Jews who made a living collecting taxes from the Jews for Rome, and they were considered traitors and cheaters. It was a lucrative job because the tax collectors would often charge more than what Rome required and keep some for themselves. As a chief tax collector, Zacchaeus was seen as a chief cheater. That is why, when he had a change of heart, Zacchaeus spoke of cheating people and paying them back.

In that culture, even more so than ours, to eat with someone was a sign of acceptance and value. When Jesus expressed a desire to eat at Zacchaeus' house, there were two different responses. First, Zacchaeus felt so honored, valued, and loved that he joyfully hurried down from the tree and had a change of heart concerning his lifestyle. Because of his repentance, Jesus declared salvation for Zacchaeus and his household. The second response was from the crowd. They grumbled because they could not understand why Jesus would be the "guest of a sinner." Why would he accept a man everyone else rejects? It was very confusing for them to see someone they admire violate the "we hate tax collectors" rule they held dear. Jesus answered that question in verse 10: "For the Son of Man came to seek and to save the lost."

People hated Zacchaeus because of his life choices and his lifestyle. Jesus never said, "Hey, Zack, I've been meaning to talk to you about your crooked tax collecting practices." This had to be quite a lesson for Jesus' Jewish disciples, one of whom (Matthew) was a tax collector! Jesus showed them that they cannot let worldly values shape their views of people because they could miss or skip over the "Zacchaeus" opportunities. Jesus disagreed with Zacchaeus' lifestyle but approached

Zacchaeus without prejudice and judgment. He was confident in the power of what his love could do. Are you?

The same Savior is doing the same seeking through us, so we must be prepared to respond in the same way. If we buy into the world's value system, it could sabotage our efforts to share Jesus with a person or people group God is leading us to. Is there someone or some group God wants to change your heart toward, as he did with Peter? Are you prepared to cross cultural barriers as Jesus did with the Samaritan woman?

When you begin with prayer, you can be sure that prayer will prepare you, change you, and direct you. When you pray, expect God to speak.

"Speak, Lord!"

There have been moments in my life where I (John) have received a random text message from a family member, a friend, or a church member that encouraged me. Not only were the words an encouragement, but the timing of the message was also significant. It would often amaze me how the message applied to what I felt at that moment. God was speaking to me through them, encouraging me and reminding me he is with me.

I have also been on the sending end of these encounters. There have been moments when someone crossed my mind out of nowhere. I did not think much about it, only to find out later that something significant was happening and, in fact, God had put them on my heart so I could pray. Truthfully, I failed to pray on numerous occasions. It was actually the constant regret of missing these opportunities that taught me to start taking them seriously.

Hearing the stories of others impacted by these situations increased my desire to be more sensitive to these moments. I'm becoming more aware of the random times people pop into my head or their face comes to mind for apparently no reason. These are the moments I have learned to recognize when God might be trying to tell me something. It's as subtle as a random, passing thought. Instead of ignoring those moments and wondering, "Where did that come from?" Ask God, "Is that you? If so, what do you want me to do?"

"Who Shall I Text, and What Shall I Say?"

A few years ago, I taught a sermon series on how to hear from God. We focused on how easy it can be to receive guidance about people we should reach out to.

In the middle of my message, I did an exercise where I told everyone to get out their cell phones. First of all, I was making the point that God can and will actually speak to you about who to talk with. Second, I wanted people to see how easy it could be to receive a response from him. I wanted them to experience getting a sense of someone God wanted them to share a message with or to encourage. In this case, the goal was to just send an encouraging message. So I told everyone, with their phone in hand, to pray a simple prayer that went something like this: "Lord, show me who you want me to text."

> Ask God, "Is that you? If so, what do you want me to do?"

After everyone repeated that prayer out loud, I said, "Now, let's pause for a moment to wait and listen. A name might come to mind, a person's face may pop up in your head, maybe a memory or a conversation with someone may come up. Just be open to however God wants to respond."

After waiting for about 20 seconds, I asked those who had identified someone to raise their hands. I had no idea how many people would respond positively. To my surprise, half of the 300 people in the room raised their hands. I said, "Great! Let's do it one more time and see if more people can identify someone." So I gave them another 20 seconds, and even more hands were raised. Next I said, "OK, now ask God what he wants you to tell that person, and send the text." I told everyone to turn their phones from vibrate to sound so we could hear people responding throughout the rest of my message. It was pretty cool to hear text notifications during the rest of the service.

The best part was all the people who came up afterward to share the responses they received. Quite a few texts were turned into Facebook posts. Not everyone received responses that evening. Some responses came the next day or days later and were posted to Facebook. The responses were variations of:

- "I can't believe you just texted me! I've been meaning to call you, but I didn't want to bother you—and here you are texting me."
- "God really loves me. This text was right on time! Thank you so much!"
- "You just don't know how much I needed to hear that. Thanks for being obedient! God is letting me know he has my back!"

This exercise could help you jumpstart your confidence in recognizing "God nudges," those moments where you feel a divine unction or impulse to reach out to someone. It will especially help you as we discuss "listening digitally" in the next chapter.

Prayer as a Constant Conversation with Jesus

As we conclude this chapter, we want to remind you that although we "Begin with Prayer," we also continue with prayer. Prayer is not something we check off a list before moving on to something else. Prayer undergirds this entire framework, and it should be central to our spiritual lives.

HeEnables, a virtual reality bi-vocational pastor of Oasis Church VR (AltspaceVR) is passionate about prayer in virtual reality, how prayer is not only the *first* thing, but it's everything:

STORY

For me, prayer isn't complex, as it's relationship with Jesus. If I know something is coming up (like an event or service), then I pray over it and ask for him to use the space we believe he told us to create. In the other situations that are in the moment, I ask what to speak, what to say, and what to see. For example, we had one woman come over to Oasis Church VR in AltspaceVR and ask for prayer for healing. She had other Christians around her from another church, and everyone assumed that she had received and believed in Jesus.

While I was praying for her, the Holy Spirit interrupted me and revealed some questions to ask her. As it turned out, she wasn't a believer. She eventually prayed to receive him right then and there. Then we continued from there to pray for healing. My belief is that prayer is a constant state of conversation with Jesus, and in these moments asking what he wants to do and doing it.

I believe that when we see more clearly how willing the Spirit is to lead us in evangelism, we will step out and take more risks. Also, when we have a better understanding of our role (including what it is and what it is not), perceived expectations will not hinder us as much. Instead, we can lean into opportunities the Lord brings our way or leads us to with more trust in his presence and in his work in the hearts of those with whom we share the gospel. Always remember that you are responsible for obedience and God is responsible for results. God uses our efforts, not our intentions.

When we begin with prayer, we remind ourselves that God owns this entire evangelism process (a theme repeated throughout this book). When we begin with prayer, we posture our hearts in submission to God's desires, plans, methods, and timing. When we begin with prayer, it helps purge us of our personal preferences and prejudices. When we begin with prayer, it prepares our hearts to listen—the next way to B.L.E.S.S. people online.

You are responsible for obedience and God is responsible for results. God uses our efforts, not our intentions.

ACTION STEPS: Begin with Prayer

Grab something to write on and to write with, and let's get ready to pray!

Prayer Prepares You

» Pray for your heart. Pray for boldness if you need it. If your heart has grown cold or numb toward the lost, pray for a heart for the harvest.

Prayer Focuses You

- » Who do you have a burden for? Who do you have a natural interest in? Is there a group you came out of that God is sending you back to with the gospel? Ask God to show you who and where.
- » Don't overthink it or second-guess it. Pay attention to the images, words, faces, and memories that pop up.
- » Write down what you hear God saying, then share it with people who know you, and get their thoughts as well.

Prayer Challenges You

- » Is there a group of people you would not want God to send you to–any "no, not them" people? Is there a group of people that you have a judgment against and are overly critical? Ask God to reveal these biases to you.

Prayer and Text Challenge

- » Ask God whether there is someone he wants you to send a message to or to pray for. Wait 20-30 seconds. When you get a name, ask him what you should say or pray. Then do it.

CHAPTER 4

To B.L.E.S.S., Move from Social to Active Listening

DAVE SAYS

Any relationship starts with listening to someone's words and life. Really hearing and understanding what someone you've been praying for is saying and feeling may be the kindest and most loving gift you can give.[38]

NOW THAT YOU HAVE AN IDEA of who God is sending you to, it's time to let them know you care. When people know you care about them, it builds trust. The best way to accomplish this is by listening to them as they share aspects of their lives online. In addition to having a specific target group or person the Spirit has led you to focus on, remember to always have your spiritual antenna up and ready for God to highlight someone to you. Barbara Carneiro, founder and senior brand strategist at Word Revolution, recently shared this digital encounter with us:

STORY

This lady shared anonymously in a Facebook group that she was having a hard time accepting Jesus and was struggling with New Age stuff she had been dealing with. A huge struggle for her because she had a great fear of hell. We have been talking since. Even though the post was anonymous, I invited her to reach out, and she did via DM. We started chatting and

> moved to text messages. She has seen a breakthrough, and although she is still exploring her faith, she has more peace about who God is. Most of her views have been on hell and sin and rarely on the good news of the gospel. Next week, we will meet in person.

Barbara is not a pastor. She's not a paid church staff member. She's an everyday believer who was socially listening. This is something you can easily do as well.

How Do We Listen Digitally?

As you grow as a digital missionary, you will want to develop the skill of listening. Tools like social media are designed for conversation. Whether talking on webcam, audio only, texting, direct messaging, or online chat room/direct message, there are plenty of ways to use this technology to listen. In digital and metaverse spaces, we let people know we are listening by the ways we respond. When someone creates content, the only way they know that someone is listening is by reactions, comments, and questions. There are a lot of spectators online. They read and listen but do not engage.

Digital missionaries engage by responding to content. We let people know we are digitally listening by digitally responding. People post so much about their lives online that there are more opportunities to respond to people than we know what to do with. In her book *Online Jesus*, Angela Craig said, "This is one of the biggest reasons social networks thrive. We post in the hopes someone will tell us [that] our life matters. We post to be seen. There is more to a selfie than showing our friends and family who we are hanging out with or where we have been. Selfies are more than a form of self-expression. They are a call for recognition."[39] When people post these "calls for recognition," digital missionaries are listening and letting them know we are listening. We can do this with quick reactions, making comments, or asking questions to promote further dialogue.

> Develop the skill of listening.

The best reactions and responses come after moving from social listening to active listening. The transition point between these two steps is when you engage with a question. Let me explain. Social listening is scanning and scrolling, looking for what might interest you. Then you ask a question (or make a comment) that introduces you to the discussion. Active listening begins when you focus on the responses in your digital conversation. Let's take a moment to unpack these three steps. Afterward, I (John) will highlight certain topics to pay attention to when you're social listening online.

Social Listening

"Social listening, also referred to as social media listening, is the process of identifying and assessing what is being said about a company, individual, product or brand on the internet."[40] Businesses conduct social listening to find out what people say about them online. When a customer shares a glowing or seething review, they want to know and engage. Social listening is used to understand online conversations about their brand. It gets much deeper than that for them, but we'll keep it simple for our purposes. You already do social listening. You always have a list of keywords in your head as you scroll through social media or online. This is how you know when to stop scrolling to check out a conversation.

Whenever you see someone mention something that represents your interests and experiences, you check it out. Suppose someone writes a post about your favorite sports team or player. It grabs your attention, and you explore the comments in the conversation. It's the same with any interest. In the same way a business wants to know what is said about them online, you want to know what people are saying about your topics of interest too. With a quick assessment, you determine whether the conversation is worth reading or joining, or whether you want to move on and keep searching. When you scroll online, you are looking for places where you want to spend more time.

As a digital missionary, you are learning to use some of your time online to be mission-focused. Join some conversations, chats, etc. These are easy on-ramps into new relationships. Certainly, if someone comments about Jesus, the Church, or God, it usually stands out to you as well. But instead of just scrolling past these, use these moments to ask God if and how he wants you to respond. There are plenty of opportunities for social listening, but we do it in four main contexts: written, audio, video, and virtual platforms. As we look at each one, I'll share some tips to keep in mind as you use it.

Written Communication (Texts, Email, Chats, DMs)

There are plenty of ways to start and nurture digital relationships through written communication.

In case you have not noticed, people like talking about themselves. Social media has made revealing private matters online more culturally acceptable. Some people might not be as hard of a nut to crack as they might have been five years ago. And, as we'll talk about later, sometimes life events make people more open to someone asking about them. Here's where good questions play a role. Initiating conversations with good questions or responding with thoughtful comments is part of what lets a person know that you, on the other side of the screen, are actually listening to them.

Although people are clearly willing to share a lot about their lives online, be mindful that as your relationship with someone progresses and gets more personal, you may want to move from a public comment section to a private chat. You can reach out to them in a direct message or invite them to reach out to you if they want to talk further. You may also transition from written to audio or video chat. You will notice this pattern in several stories in this book.

One benefit of written communication is that chats can go on for days as people comment and add to the conversation. This means you can have time to pray and process how you respond. Unlike audio and video communication, you can take hours to respond or even wait until the next day. When talking with someone about faith, you have time to research ideas and find answers you don't have. Also, you can share other online content for people to check out when the time comes.

One challenge of written communication is that things like sarcasm and humor are sometimes lost because we cannot hear a tone of voice or see facial expressions. Be mindful that written communication requires a bit more self-awareness in your choice of words. Also, it requires social awareness as you respond to what others say. When you wonder what a person means, it's best to ask for clarification—especially as you get to know someone.

Audio (FB Audio, Twitter Spaces, Clubhouse)

Advances in audio communications have been impressive. I (John) remember it being a big deal when three-way calling was introduced mainstream. Now there are apps that let you be part of an audio conversation with thousands of people. Using an audio app or audio feature to communicate in some platforms enables you to have conversations with people while still keeping your phone number private. This gives both parties a sense of privacy early in a relationship.

Just a word of caution. With audio only, we are often busy doing something else because we are not seen. But multitasking can be a problem when listening to someone's story. As much as possible, give your undivided attention to these people God has brought your way. If not, you may miss important personal details they have shared or a window in the conversation when you could ask a good follow-up question or present the gospel.

People have a way of knowing when you are distracted on an audio call or whether you are actually paying attention. If they share something serious and your response is silly, they know you were not listening. Or if they share something serious that they feel you should have responded to and you ignore it, they'll also feel like you were not listening. Be fully present in these conversations as you build trusted relationships online.

Video Chats

Listening via video chats has the most advantages and is the most like being in person.

When you are listening to a person share their life via video chat, facial expressions reveal information you don't have access to in audio or written communication. As you pay attention to facial expressions and body language, these give clues about what kind of questions you can ask. "I noticed when you said that you kind of smiled. That must have been a fond memory. Tell me more about that." Or "You say you're fine, but you don't look yourself today. What's going on?"

Seeing someone face-to-face also adds to the level of connectedness felt during the conversation. Unless you're on Zoom with a recording feature, most platforms with video chat do not record. This also creates another level of safety and privacy for transparent discussions. Use your own discernment to determine whether there is enough relational equity for you both to be comfortable moving from a written or audio-only conversation to a video chat.

Virtual Reality

Social media is incredibly conversational. As anyone who's ever been in virtual reality can tell you, VR is entirely conversational as well. Building relationships with people in virtual worlds like AltspaceVR, VRChat, RecRoom, or Meta Horizon Worlds is very easy. These VR communities are designed to get strangers to talk with each other. Literally, there are worlds where people sit around a campfire and just talk to each other in a non-judgmental, non-combative environment. Using virtual reality to listen to people is easy when we position ourselves humbly.

Recently I (Jeff) was talking with a virtual reality pastor who was lamenting how difficult he perceived talking with people about Jesus was. He described scenarios where he was talking to people in virtual reality, but because the person was in avatar form, the pastor could not read body language or look at facial signals (like whether the person was smiling or not). The opportunity in virtual reality is not in pushing the idea of Christ but in taking the time to listen. Giving someone permission up front to talk gives *you* the opportunity to not only listen to their story but to analyze how they communicate and how interested they are in developing a relationship.

We need to realize that virtual reality, as a platform, is designed to get people to meet each other. VR experiences are created for users to connect and talk to each other, regardless of where they live. Truly, this technology was created to get you to listen to others.

So while you're exploring virtual reality, start talking with people! As you're walking around AltspaceVR, try out a Campfire world and sit next to an avatar. When you're playing Population One, take the time to meet and talk with your team. Even something as nefarious as sitting next to a stranger at a virtual reality bar in VRChat can be used for the Kingdom if we are using our headset's microphone to invite conversation and our headphones to listen.

Most virtual reality users love to talk. As metaverse missionaries, this is our opportunity to engage them.

The Power of Asking Good Questions

Recently I (Jeff) was on a Zoom call with Jeff Vanderstelt, director of missional communities at Doxa Church in Bellevue, Washington. With a passing (and completely off-topic) thought, Jeff blew my mind:

> In the past we could assume some starting-point assumptions like people generally believed in God; there is a universally accepted morality; people will be judged one day for their actions; there is an afterlife; etc. However, most of those can no longer be assumed. One thing people do believe is that we are relational beings longing for significant relationships. In light of this, our evangelistic work must shift from merely addressing the rational mind to the emotional and relational needs. People become aware of their relational needs by feeling their emotions and giving voice to the needs their feelings awaken. This shift requires that we become good at asking questions that awaken the heart and listening carefully for the relational needs that are revealed. Then, sharing how Jesus can truly fulfill what they

long for. This is not new. Jesus was a master at this. Which is why he said things like "Blessed are they who mourn for they will be comforted." He regularly addressed relational needs through emotional means. And he was a master of listening to the heart.

The shift from addressing rational minds to relational needs requires us to ask good questions, ones that awaken the heart and reveal relational needs. Notice that two components of B.L.E.S.S. (listening and serving—which both meet relational needs) are preceded by asking good questions. This actually works to our benefit as digital and metaverse missionaries because for some people it's actually *easier* to have spiritual conversations digitally or virtually. These technological tools can be very relational, allowing people to process their emotions authentically. It's great to engage in small talk, but it's even better when the right questions take conversations deeper. Even God uses questions to initiate relational connections.

> VR users love to talk. As metaverse missionaries, this is our opportunity to engage them.

Why Does an All-Knowing God Ask Questions?

Throughout Scripture God initiates conversations with people by asking them questions. After Adam and Eve ate from the tree of the knowledge of good and evil, God asked Adam, "Where are you?" (Genesis 3:9, ESV). He did not ask to gain information but to pursue a connection. In another instance, when God sought to partner with Ezekiel and include him in the process of restoration of Israel, God asked Ezekiel, "Son of man, can these bones live?" (Ezekiel 37:3, ESV). After his resurrection, when Jesus sought to reconnect with Mary at the tomb, he asked, "Why are you weeping?" (John 20:15, ESV). Also after the resurrection, when Jesus was walking with the disciples on the road to Emmaus, he asked them, "What are you discussing?" (Luke 24:17, NIV).

Asking questions is about building bridges and strengthening relationships. God is all-knowing, and yet he asks questions. He does it so we have the opportunity to share. For God, asking questions is not about gaining knowledge. It's about building connections. As we engage people online, not all of our questions are to gain knowledge.

Asking questions:

- » initiates conversations;
- » invites people to reveal their needs;
- » shows people we care beyond an emoji;
- » makes people feel listened to;
- » shows your interest in their lives; and
- » means you're choosing to spend time with them online.

You might say, "But this is elementary and so simple." True, it is simple and meaningful, and it is a reflection of the way God reaches out to us. In getting to know an unsaved person, you are building a relationship. You are demonstrating a side of Christianity that is perhaps different from what they have seen on TV, experienced in person, etc. You have the opportunity to show the love of Jesus, who is seeking to save those who are lost through you.

> God initiates conversations with people by asking them questions.

Asking with Curiosity

Curiosity is key to learning to ask good questions. It prevents us from making false assumptions about people and drawing inaccurate conclusions about their lives. Lead with curiosity. Ask questions and find out why, how, and when. Lean into the details of people's lives when they are sharing with you online. Each thing people share is a door into another room of their life. Asking questions about those details knocks on that door and allows people to open it or leave it closed. If people want to leave the door closed, they will definitely let you know. But if they want to open it, they will. And when they do, you will be ready to listen. Your curiosity communicates value and investment. When people feel valued and seen by you, it builds trust and connection.

To ask good questions, it helps to be genuinely interested in and curious about another person's life. This is something I (John) had to learn. I am much better at initiating conversations and keeping them interesting than I was years ago, but it came with practice. I still admire people who can strike up a conversation with anyone and within minutes, it's like they grew up together. That is definitely not me. But I've learned that it doesn't *have* to be me. It doesn't have to be *you*. You can increase your ability to interact without being as good at it as those who do it intuitively. I remember asking God, "Why do I find it hard to have conversations with strangers?" Without delay, almost as though he was waiting for me to ask, he responded, "Because you are not interested in them." Wow. I never realized it until he said that, but I could see it afterward.

Practicing Curiosity

Usually, when I am out and about, I am busy and not interested in being interrupted or delayed by small talk. I had to practice being curious. I wanted to get better at striking up meaningful conversations with people I did not know. I knew that I wanted to be able to talk with more people about Jesus. But first, I had to just talk to more people—period. I had to get over that hump, especially if I wanted more relationships with non-Christians. I spent a lot of time at coffee shops writing, having meetings, making phone calls, etc. So even when I was busy, I forced myself to start asking questions. I asked questions about the stickers on laptops, the meaning of tattoos, the book someone was reading, or what couples thought was the key to lasting relationships. It was my version of in-person social listening!

In doing so, I learned several things. For instance, most people are happy to talk to someone who shows interest in something about them or something they are interested in. I also realized that I was in fact naturally curious. I often wondered about things I observed about people. I was just not curious enough to actually ask the questions. Honestly, as I thought about this I realized it was because I did not want to make time for the conversations. But that's changed now, and

I'm grateful to God for it. Practicing curiosity in person has made me better at striking up conversations online.

When we're busy with our own lives, we tend not to be curious about the lives of others. Let's slow down, tune in to what God may be saying to us or drawing our attention to, get curious, ask questions, and listen.

The Most Important Person in the World

I (John) remember hearing a journalist describe the difference in her experience interviewing two global leaders. In her interview with the first leader, everything went as expected. She asked questions about his life, and he answered them. She left his office feeling like she just had the opportunity to talk with the most important person in the world. The second leader she interviewed asked her questions about her life while he listened. When she left the office of the second leader, she felt like she was the most important person in the world. Listening has the power to change how people feel. And because it can change how people feel, it can change their lives. This is the power of active listening.

> Slow down, tune in to God, get curious, ask questions, and listen.

When you are curious, you will have no problem asking good questions because really good questions come from a genuine interest in and love for people and a desire to get to know them by learning their story. We listen with a curiosity about the struggles that they have endured or the obstacles and challenges they have had to overcome. Then we ask questions such as these:

- What was that like for you?
- How did it impact you?
- What did you start to believe about yourself or the world or God?
- How did you come back from it and move forward?
- In what way is that experience still affecting you today?
- What questions do you have for God in light of those experiences?

And then we truly and intentionally listen to the responses.

One final thought on curiosity is how it can help to overcome a judgmental attitude. If there are things you find yourself wanting to criticize, turn the criticism into curiosity: "Hey, I'm curious about what you said concerning _____. What did you mean?" or "Could you say more about that?" This is especially helpful when you want to respond to something that offended you. If you are judgmental about a person's way of life, you cannot be genuinely compassionate. When we shift from judgment to curiosity, our compassion can be activated.

> Turn criticism into curiosity to activate compassion.

DAVE SAYS

Learn to be a great listener.

Some of the B.L.E.S.S. practices will come more naturally than others. One of the most difficult skills to develop is listening. There are at least four types of noise that make it hard for us to listen to others: 1) physical noise that happens in the setting, 2) psychological noise that happens in our minds, 3) physiological noise that happens because we are feeling something in our bodies, and 4) semantic noise because we get confused by the meaning of the speaker's words.

To really become a great listener requires us to block out those distractions. And when there are semantic misunderstandings, we ask the person to clarify them for us. You can become a great listener, but it requires a lot more intention than most of us think.[41]

The Power of Active Listening

Have you ever experienced the frustration of talking to someone for a while only to realize that they were not paying attention to you? If so, you might have responded with, "Are you listening? Did you hear anything I said?" It is natural to feel disrespected or devalued when that happens. When someone is not listening to us we can take it personally, especially if we are vulnerable and sharing something very private. The opposite is also true. When you take time to actively listen to people share their life with you, it has a positive effect on how they feel. This is the power of active listening. I (John) call it the "power" of active listening because of its influence and impact on people. These are not just benefits of active listening. It's powerful. Let me explain.

The messages people receive when we listen are:

- You are valuable.
- You are worth my time.
- Your story and experience are valid.
- You can trust me.
- I am safe.

In a world where people are more isolated and lonely than they admit, active listening creates connection. Listening is comforting in a world where more people are depressed and struggling with anxiety than ever before. That is why it is so powerful. Let's define active listening, and you'll see what I mean.

What Is Active Listening?

First, active listening is listening with undivided attention. It's not allowing yourself to be distracted. Nowadays, when we are on a Zoom call with our video off or on an audio call, we tend to multitask. Active listening requires listening without multitasking.

Active listening is also listening without judgment. Being judgmental is repelling. It makes people defensive and guarded, and it breaks down trust and connection. It is a sure and quick way to shut down a conversation, especially one about faith. On the other hand, active listening seeks to create a safe space for people to share without being judged. When you listen in this way, you become a person others can trust. Trust is just as important in digital and metaverse spaces as in physical relationships—sometimes even *more* important because of the depth of what people are willing to share online. In the past few years, we have seen more than ever before how being judgmental damages online communication and relationships. You can capitalize on this by offering a nonjudgmental approach that is refreshing and honoring, sets you apart from the crowd, and creates a relational bond. You can create a safe space for people—especially people whose life choices differ from yours.

Active listening also includes giving people meaningful feedback. This is true in digital communication, especially in the context of written communication like comment sections, chats, or direct messages because you cannot see a person's face. The quality of their comments or follow-up questions indicates how well a person is listening. You will quickly see the results when you do these things because people appreciate it so much that they often comment about it.

The Ministry of Active Listening

While serving as an Air Force chaplain, I (John) did a lot of listening . . . and I mean a *lot*. When I started active duty in 2002, a mentor told me that the amount of counseling military chaplains do in three years is equivalent to ten years as a civilian pastor. After serving nine years as a chaplain and eight years as a civilian pastor, I couldn't agree more. After hundreds of hours of counseling, I can confidently say that listening well is a ministry in itself. I cannot tell you how many people came to my office, talked for an entire hour, and said, "I needed that. Thanks for listening."

So many times, I listened to situations that I wished I could fix but couldn't. I wished I could solve the problem for them, but I couldn't. I wished I could rescue them from their dilemma, but I couldn't. I wished I could heal their pain, but I couldn't. There were situations where I felt helpless, even as a trained professional. I often wanted to do more—until I learned the ministry of active listening. I learned that these people felt loved, comforted, and validated when I listened. I learned that listening was not only doing something, but in many cases, listening was exactly what was needed.

DAVE SAYS

In 1816, René Laennec, a French doctor who was uncomfortable about placing his ear directly on a woman's chest to listen to her heart, invented a tool that doctors continue using today: the stethoscope. He was famous for telling doctors and nurses, "Don't stop listening to your patients–they will still tell you how to heal them." Laennec didn't want doctors to just listen to hearts and lungs with his invention. He wanted to make sure they still listened to the words and feelings of their patients.

I wonder if we understand that we can be a huge gift to others just by listening to their hearts. A writer friend once told me, "Dave, everyone has a story, but [they have] few people who will listen to that story." One of the most practical and powerful ways we can be blessings is to simply listen. Listen to people. Listen to our community. Listen to our coworkers. Listen to our classmates. Listen to our neighbors. Listen to the hurts, the needs, and the pain of the people in our lives. Listen to their stories, and discover what they really need. When was the last time you asked the single mother who lives near you how she's doing? Or stopped what you were doing to give your undivided attention to the person in the cubicle next to you? When was the last time you slowed down enough to listen to the cashier who casually mentioned that her husband was hospitalized?

When I say, "Listen to people," I don't just mean listen to the words that come out of their mouths. I'm talking about listening to their hearts! Over 20 years ago, Stephen Covey wrote his landmark self-help book *The 7 Habits of Highly Effective People*. People have bought his book by the millions, including me. And one of the habits I'll never forget is this: "Seek first to understand before seeking to be understood."[42]

Jesus as an Active Listener

We often talk about Jesus as a storyteller. Still, one of the things that I value more and more about the ministry of Jesus is his intentionality in asking questions and listening to people's responses and their stories. This says a lot about his heart for people. Jesus—the Son of God, the Savior of the world, and the King of kings—made time to listen to people. I (John) like Dave's observations:

DAVE SAYS

Luke's account of Jesus and the blind beggar is a great example of Jesus as an active listener:

When Jesus approached Jericho, a blind beggar was sitting beside the road. When he heard the noise of a crowd going past, he asked what was happening. They told him that Jesus the Nazarene was going by. So he began shouting, "Jesus, Son of David, have mercy on me!" "Be quiet!" the people in front yelled at him. But he only shouted louder, "Son of David, have mercy on me!" When Jesus heard him, he stopped and ordered that the man be brought to him (Luke 18:35-44, NLT).

Pause a minute with me, because I don't want you to miss this. Jesus wasn't necessarily looking for someone to heal that day. He was on his way to Jericho, one of the last stops on the way to Jerusalem to participate in the Passover Feast. Jericho was extra-busy that time of year. Picture Times Square on New Year's Eve. Hordes of people. And in the midst of them was a blind man crying out in all the noise, "Have mercy on me!" In all that commotion, Jesus heard that singular voice. It makes me think of how my wife, Sue, could hear even the faintest cry of one of our kids at night when I was lying next to her, oblivious. Jesus heard the blind man the same way. The people traveling with him told the man to leave Jesus alone. But he wouldn't. Instead, he cried out even louder.

And here's what happened next: "Jesus asked [the man], 'What do you want me to do for you?' 'Lord, I want to see!' he answered. Jesus replied, 'Look and you will see! Your eyes are healed because of your faith'" (Luke 18:40-42, CEV). You knew that was coming, right? Blind man can't see, but he calls out to Jesus, and Jesus heals him. The end. But not so fast! Notice the question Jesus asks: "What do you want me to do for you?" Seems like a silly question, doesn't it? I'm sure Jesus knew the man was blind, so why did he ask that? Is this some sort of dramatic buildup for the miracle to follow? I don't think so.

I believe something is happening here that we could easily miss in a quick read. The question Jesus asks tells us something profound about his character. It shows that Jesus didn't assume he knew what people needed—so he asked questions and listened. You want to bless your neighbors, friends, and family? Imitate Jesus in the way you listen to them. It's a challenge not just to hear the words they say, but to really listen.[43]

As digital missionaries, we have a responsibility to understand the people we are talking to, which is why (if we are going to B.L.E.S.S. people digitally and in the metaverse) we must listen far more than we speak. The great American theologian Francis Schaeffer is quoted as saying, "If I have only an hour with someone, I will spend the first 55 minutes asking them questions and finding out what is troubling their heart and mind, and then in the last five minutes I will share something of the truth."[44] At a philosophical level, we cannot live at peace with people unless and until we understand them. And if we don't yet understand them, we can begin by listening. That's what we love about this next story.

Love Them by Listening

Anthony English, a Christian blogger, shared this story earlier this year in a blog entitled, "The Three E's of Evangelism:"

STORY

Recently, I was speaking with a young man at a local park. Through the course of the conversation, he had mentioned his rough upbringing and how it shaped his view of the things of God. I could've gone straight to answering some of his objections, but I didn't. Instead, I asked him to share more of his story. As he did, his tears started to flow. He poured out his heart and allowed me to share in his pain and sorrow. When he finished, I was able to minister to him further by offering some perspective on his trials and questions that he had. At the end of our conversation, he mentioned how he needed to rethink his views on God, and I was able to share the gospel of Jesus with him. It was a good time of fellowship and encouragement for us both, and I wonder if one of the reasons for that was my willingness to simply love him by listening.[45]

Anthony's story is a great example of the kind of impact we can have on people when we listen. People feel loved when we listen. Evangelism is most effective when people experience God's love for them. Listening is one way that can happen.

A Few Things to Listen For

There are two specific types of comments to pay special attention to. One is the "cry for spiritual help" post or comment. The other is when people share a significant life event. I (John) am not saying that you are obligated to respond every time you see one of these comments. That would be too overwhelming. However, I suggest that you at least take a moment and ask God whether you should respond and, if so, how. Scrolling past these comments can make us numb to the pain and suffering behind the post. We cannot do things that make us numb to pain and then turn on compassion when we want to use it. No one was more loving and compassionate than Jesus, yet he only did what the Father told him to do (John 5:19). Jesus' ministry was not driven by people's needs but by the Father's voice. The same goes for you.

Jesus' ministry was not driven by people's needs but by the Father's voice.

Cries for Help

People ask for help all the time online. They love to tell their story to someone who will listen. Sometimes these cries for help are posted hoping for someone to respond with attention and compassion. But I want to focus on a *specific* type of cry for help. Often, when people share some kind of desperation about their relationship with God online, they are inviting people to follow up with them.

"I Feel Lost. I Need to Find My Faith Again!"

Church Anywhere is a digital church extending from First Capital Christian Church in Corydon, Indiana. One of their online pastors, Megan Carter, who we mentioned earlier, shared this story with us. It illustrates a great way to respond to a "cry for help" post.

STORY

February of 2021 was the first time I connected with Jess. She messaged me because I commented on a different person's TikTok. She had been following my page for a while, and after that initial connection, we began chatting about her faith. In her words "I feel lost, and I also feel like I need to find my faith again." I was able to talk with her about her faith, and we also talked a lot about mental health. We met on a Zoom call and talked about her life situations. They were really heavy, but I was able to encourage her and give her hope and pray with her. I sent her a Bible because she didn't have one, and she started reading it. I love getting messages from her asking questions about what she is reading so she can understand it better. She's learning, she's growing, and she knows I'm always available to pray for her whenever she needs me.

Let's highlight a few takeaways from this story. First, as a digital missionary, Megan was commenting on someone else's TikTok page. Her comment on someone else's page motivated Jess to reach out. But what have we talked about? This was not just Jess reaching out. This was God at work. Because Megan made herself visible and available on TikTok, she received a referral from God. Right? That's exactly what happened. We need to be actively engaged in conversations and sharing insightful feedback instead of only reading without engaging. Second, Jess had been observing Megan online for a while. Megan's digital reputation, which we will discuss later, was part of Jess' decision to reach out. Jess trusted Megan. Next, when Jess made a "cry for help" post, Megan followed up and moved to a private conversation on Zoom, where she could listen more to her needs. Megan was able to serve Jess by getting her a Bible. Serving is another practice we will talk about later.

Significant Life Experiences Open and Soften Hearts

When we listen to people in digital spaces, we want to hear more than information. We want to know their story. We listen for not only the events of someone's life, but how those events shaped who they are today, their worldview, things they are proud of, things they are afraid of, and so on.

When I (John) was in a class at Gateway Seminary (formerly Golden Gate Seminary), president Dr. Jeff Lorg spoke to our cohort about evangelism. In his lecture, he shared about how people are often open to hearing the gospel when they have experienced a significant life event. Significant life events can be positive or negative. Good experiences like marriage, physical healing, or the birth of a child can cause people to re-evaluate their lives and provoke a search for meaning. These experiences are often turning points where people are open to hearing the gospel of Jesus. When people are experiencing pain, they are often open to hearing words of hope and encouragement. They look for a way that their lives can get better, more stable, or back on track. When people experience the death of someone they know, a divorce, or a severe illness, these situations can create a search for answers, hope, comfort, and curiosity about the afterlife.

You may have experienced some of the life events I describe below. Keep in mind that I am not trying to create an exhaustive list of every possible feeling or reaction. I simply want to give brief explanations for the benefit of those who have not had these experiences. My goal is to help you look beyond the surface and better understand how to engage someone else.

Death of a Loved One

When people experience the death of a loved one, it opens their hearts to sources of hope and encouragement. When someone we love or know dies, it causes us to examine our beliefs about mortality and the afterlife. Sometimes people put off seeking the answers to some of life's most essential questions, thinking that they have plenty of time to figure it out later. However, when someone we know dies, especially someone we love, these crucial questions often become urgent. This dynamic is true when people experience a recent death, but the grief and the questions that grief can present may last for years. As people go through different stages of grief at varying levels of intensity, this can lead to conversations about the gospel of Jesus.

Divorce

Another experience that can lead to openness to the gospel is divorce. A divorce is a kind of death as well. People experience the death of the dreams of the life for which they had hoped. Trust, hope, and security, among other things, have died. Divorce can leave people feeling unstable and unsure about the future. The relationship they thought they could trust and count on has been damaged, destroyed, and ended. The heartbreak experienced in divorce can often leave people open to hearing the good news of Jesus. When a person's world is turning upside down, they look for stability, certainty, truth, hope, and healing.

Sickness and Hospitalization

As a former hospital chaplain in both civilian and military hospitals, I can tell you firsthand that severe illness and hospitalization will cause people to think about their mortality and the afterlife, similar to when someone dies. Another experience with a similar effect is when a person is diagnosed with a disease, especially a life-threatening one. When people consider their mortality, they face questions for which they do not have answers. The need to find solutions can lead them to be open to hearing the gospel of Jesus.

People usually do not mind letting you know they are in the hospital. Sometimes they post pictures on social media of themselves in the hospital. "They're only asking about Jesus because they're sick, dying, or looking for attention"—ever thought that? The truth is, even if you are right, how many of us had life experiences that caused us to take another look at Jesus? Don't go down the road of judgment. Choose compassion, love, and encouragement. Be open to questions or interest in faith conversations, whatever the reason behind them.

Birth of a Child

As mentioned earlier, although significant life events can be adverse and painful, there are also positive significant life events that are reasons to celebrate. One experience that causes people to evaluate the direction of their life is the birth of a child. For this reason, baby dedications where one parent is Christian and the other is not can create opportunities for discussion. Sometimes individuals living for themselves and making unwise decisions decide to "get their act together" when a child is born. Now that another life depends on theirs, many people make a significant shift when they become parents. They want to make more responsible decisions and live a life that will foster a healthy relationship and environment to raise the little one. (This response is not happening as often as it was a decade ago because cultural views toward the church have changed; but it is still something to keep an eye on.)

Significant life events motivate people to ask important life questions. When they do, these are great opportunities for God to reveal himself to people looking for help, hope, support, answers, etc. So as you listen to people online, pay attention when they share a significant life experience. Sometimes when people open up to you, it is their way of looking up to God.

Remember, these life experiences do not necessarily have to be recent. An experience can be a few years old, but the individual may still wrestle with thoughts, ideas, and questions. So if someone talks about a significant life event they had years earlier, it's still an opportunity for you today to listen and to ask meaningful questions.

DAVE SAYS

Listening moves us forward in our relationships as God uses us to bless others and change their lives. It's the next step in loving your neighbors and helping them know the love of God.[46]

We have talked about beginning with prayer and about listening. The next part of the B.L.E.S.S. framework is Enjoying Shared Experiences.

ACTION STEPS: Listen

Social Listening Exercise #1

ASK WITH CURIOSITY: If you want to be more comfortable initiating conversations with people, start by asking God what is holding you back.

Social Listening Exercise #2

Take a break from reading this book and log into your favorite online platform. Set a timer for 10 minutes, and take some time for social listening. Scan and scroll!

- Pay attention to what grabs your attention and why.
- What conversations do you feel drawn to?
- See whether you notice any cries for help. If you do, pray for that person, and ask God whether and how you should respond.

Examples of Good Questions:

- Can you say more about that?
- How long have you been feeling this way?
- Who around you can help?
- What has caused you to feel that way about God?
- I'm sorry to hear that. How can I pray for you?

CHAPTER 5

To B.L.E.S.S. Digitally, Enjoy Shared Experiences

DAVE SAYS

Of course, we know Jesus was neither a drunkard nor a glutton, but he so frequently ate with people who were that he was accused of it—a lot. Because for Jesus, eating was essential to his mission of seeking and saving the lost. He blessed and loved people by sharing meals with them.[47]

Jesus and Shared Experience

IN DAVE'S BOOK, THE E IN B.L.E.S.S. IS "Eating with Others" because of the bonding that happens when sharing a meal. Certainly, Jesus did that quite often. However, there were also times when he invited his disciples to share an experience with him. Right from the beginning, Jesus invited his disciples to leave their old lives behind and do life with him. He saw shared experiences as a core part of discipleship. Then there were more specific experiences, such as Peter being invited to walk on water with Jesus. That was a moment only the two of them shared. At Peter's request, Jesus invited him to "Come!" (Matthew 14:29). Peter, James, and John were invited to experience Christ's transfiguration, the Last Supper, and prayer in the Garden of Gethsemane.

Shared spiritual experiences with Christ are core elements of the gospel. We have been crucified with Christ (Galatians 2:20), raised up with Christ, and seated with Christ (Ephesians 2:6). The ultimate goal of the gospel is this: "And if I go and prepare a place for you, I will come again and will take you to myself, that where I am you may be also" (John 14:3, ESV). Jesus wants us to share in his experience of God. In fact, God the Father, God the Son, and God the Holy Spirit exist in an eternal shared experience with each other. Now, through Christ, we get to share their experience for eternity. We were created to share experiences with God and with each other. This is why it is so meaningful when we do!

When we share experiences with people, it brings us together. There is a social bonding that happens even when it is not done intentionally and no one is aware of it. However, when we *are* aware of it, we can use it to our advantage in starting and building new relationships or strengthening old ones. Because this is so important to social connection, the opportunities to enjoy shared experiences digitally and metaversely are in abundance. When we are looking to build relationships online, we need to think strategically about those opportunities. As we connect around shared interests about nonspiritual things, we lay a relational foundation for conversations about spiritual things. As we begin with prayer, as we listen, we may find that God begins to open doors toward building relationships.

Experiences Are for Nerds

We have an incredible opportunity to engage new people through digital and metaverse communities, and the uniqueness of these communities means that we can engage people where they are. Some are using this opportunity to connect with gamers and other hard-to-reach digital communities. Others are finding ways to connect with a more typical person via digital methods.

Let's insert Mark Lutz into the conversation. Mark is a digital church planter for a gaming church called Lux Digital Church. We don't talk much about pastors or church leaders in this book, but Mark has been doing an incredible job connecting with people digitally through shared experiences like gaming. Listen to Mark's perspective.

STORY

I (Mark) have been playing *Dungeons and Dragons* for over 20 years with the same people: my friends Joe, Alex, and Hunter. A few months back, everyone was in the same area at the same time, so we got together for an evening to catch up. That night, Hunter came from his folk's place (where he was staying while in town, not because he lived in their basement) and brought a folder with him that he had found there. It contained details and papers from years of playing tabletop roleplaying games together. We went through those papers and recounted stories of past gaming sessions like you would see a few old baseball players reminiscing about their high school team or a dad and his sons sharing stories from hunting trips of the past.

Those shared experiences formed us and forged my longest-lasting and most impactful relationships outside of those in my immediate family. So we are using that tool in our church. We play *Dungeons and Dragons* together online and invite our unbelieving friends to join us. Those shared experiences of battles fought and monsters slain generate relationships that lead to deep spiritual conversations.

Recently LordShadow joined our *Dungeons and Dragons* community. He is an atheist and only there to play in the games. After a few games, we asked how we could pray for him, and he opened up about losing someone close to him from suicide. We prayed for him that week.

"Nerds" are on edge and defensive of any religion because of the way we have been treated and then dismissed. So when a nerd finds another nerd, we almost instantly connect. That connection and shared experience around the table or through a video game disarms the normal defenses and allows for deep, genuine, spiritual conversation that often results in life transformation.

Experiences Are for Everyone

I (Jeff) hear you. I can see where you, the reader, may be thinking, "*Dungeons and Dragons* is a little creepy; besides, normal people don't get obsessed with things like that." While it's somewhat true that "normal" people don't get obsessed with *Dungeons and Dragons*, there are plenty of other things they *can* get obsessed with.

The metaverse is literally defined as the "Age of Experiences;" people are all about experiences (and they're becoming even more so). You may be playing a tabletop game digitally or jumping into a virtual world with someone and walking through the experience together. Maybe the experience is watching a YouTube video and discussing it with someone or agreeing to be remote running partners. Perhaps it's working through a virtual reality escape room with someone, or maybe it's just jumping on a Zoom call or Facebook Messenger and talking via webcam—or even old-fashioned texting! The point is that experiences can be virtual, or digital, or physical. They don't have to look like eating a physical meal together, nor do they have to be video-game based—but the experience *does* have to drive the relationship closer. What creative thing can you do to drive a relationship closer?

What creative thing can you do to drive a relationship closer?

Why Do Sunday Morning Online Church Services Struggle as an Experience?

Technically, an online church service can be a shared experience! It does require, though, that the experience truly be shared.

The typical church struggles with online services because the church has no idea who is watching these services. It's very easy for online services to be consumeristic because people could be watching services online without any relational connection.

Remember, as digital missionaries, we're aiming to connect with people who are cold to Christ. We're attempting to engage them on an emotional level that is grounded in relationships. Is a 70-minute online church service the best approach at this point? Maybe.

- If a person is comfortable with church experiences and has just fallen away in recent years, then maybe a shared church experience has value.
- If a person is more "de-churched" and has been hurt by the church, then a shared experience centered around a church service could do more harm than good.
- If a person has no experience with church, this is an opportunity for a digital missionary like you to find experiences that are more on par with where that individual is, even if that takes place outside a church service.

The most powerful part of the experience is not the actual experience but the conversation you have during and after, as you build a relationship with someone who needs to know Jesus. Church services can be helpful, or they can be a block. Any time you use an online church service (or even a physical church experience, for that matter), remember to use it as a relationship-builder. Don't rush to use it because it is comfortable for you. Be discerning about when the time is right for the person you're inviting. Often, inviting someone to church happens prematurely because that is all many Christians know how to do—but not you! You have the B.L.E.S.S. principles to help you develop a deeper relationship with the person you aim to lead to Christ. This way, when you do introduce them to a physical, digital, or metaverse church, they will be more likely to stick.

Missional Imagination: Finding the Fun in Shared Digital Experiences

I (Jeff) don't think we truly appreciate how fun our God can be! That sense of fun is often lost when we communicate through traditional "church." But shared experiences give digital missionaries the opportunity to lean in and find that joy again while using their "missional imagination." Missional imagination is our opportunity to think outside the box.

Biblically we see Paul address this in Ephesians 3:20 (NIV): "Now to him who is able to do immeasurably more than all we ask or imagine, according to his power that is at work within us." God wants us to ask, to imagine, to dream. And he wants to fulfill that imaginative wish within us. Digital missionaries need to tap into their creativity, while tapping into the Holy Spirit, and find ways to connect with people digitally or in the metaverse.

The experience does not even have to be Christ-centered! Ephesians 4 beautifully explains that God can be seen in everything, even in the metaverse! Ephesians 4:4-6 (NIV) says, "There is one body and one Spirit, just as you were called to one hope when you were called; one Lord, one faith, one baptism; one God and Father of all, who is over all and through all and in all."

There are four levels of digitally shared experiences. Each has an increased level of engagement and an increased potential for deeper connection.

Levels of Experiences

Let me (John) clarify the different levels of shared experiences. All of them are great connection builders, but each of them builds a different level of connection.

Sharing Your Experiences

This is when you share photos and videos of vacations, food, celebratory moments like weddings and gender reveals, even tributes to someone who has passed away. The experience was yours, and you are reporting what happened. Social media has allowed us to share moments of our lives with the world. As we share our lives with people, we invite them into our world. There is a level of connectivity we discover with these friends—that's why we do it! If you are really into social media, sometimes you feel bad when you forget to share a post. I've said a few times, "Oh! I forgot to take a picture so I could post it!"

Sometimes I've seen a post with the words "late post" to indicate that the person did not post it when it was happening, but they still wanted to share it with their followers. Think about that. Why? Because there is a level of connectivity when we invite people into our experiences for people to comment. We want to share our joys and our pains.

Shared Life Experiences

As a United States Air Force veteran, any time I (John) see someone wearing a veteran hat or shirt, I feel a connection even before speaking to them because of the shared experience of serving in the military. I'm in a Facebook group with over 718,000 veterans. There are military jokes that you would only get if you served in the military. There is shared advice on veteran benefits, where to go for help when struggling with PTSD, tips on getting jobs when you leave active duty, shared memories, and the list goes on. Online parenting groups allow people who've never met to support each other, give advice, and share the struggles of the toddler, teen, or empty nest years.

Participating in digital and metaverse communities based on shared life experience is a great way to quickly establish communication with absolute strangers. People love talking about their passions and knowledge. These opportunities do several things. For one, they make it easier to break the ice with strangers because there is already a shared interest that you are comfortable discussing. In any interest group, you will find people at different levels of expertise. Engage at whatever level you can.

For example, a cooking group may have expert cooks, bakers, and chefs as well as newbies. If you're an experienced cook, be generous with your cooking tips and advice. You may end up mentoring someone via direct message. If you're new to cooking, be generous with your questions. People love answering questions! Also, be generous with your compliments after receiving advice that was helpful. People also love knowing that they were able to help someone. No matter what level of expertise you have in a particular area, you can usually find that sometimes you're offering your opinion or advice, and other times you're asking for opinions or advice.

Shared Audience Experiences

Think about why people gather for Super Bowl parties at someone's house. Or why when a team is playing away from home, thousands of fans will travel to the stadium and watch the game on massive screens. Why not just watch it at home and talk about it the next day? What's the importance of being in the same room with everyone? It's the value of enjoying shared experiences—even as an audience. This is what Facebook created with their watch parties.

Not too long ago I (John) flew out of town for a church consultation. I flew in on Friday evening, facilitated the consultation Saturday morning, and had the evening free because my message for Sunday was already dialed in. I decided to catch a movie (one I did not think my wife, Jeanine, would be interested in seeing together). It was a good movie, but because I was confident that Jeanine was not interested I didn't even bring it up when I returned home. We had a date night about a week later. To my surprise she said she wanted to see the same movie. I said, "Oh, I already saw that one. It's pretty good!" She looked at me like I had just denied Jesus three times. She asked, "When did you see it?" Realizing that I should have "Begun with Prayer," with less confidence I slowly said, "Uh, last week, when I was out of town. I picked a movie I thought you would not be interested in. That's why I didn't tell you—because I didn't think you'd like it anyway." Even though I knew she would say no, I still tried: "I'll watch it again with you?" To make a long story short, let's just say I received undeniable clarity that erased any further confusion.

You may have a similar story or at least know of one. Why is this such a big deal when this happens with our friends or family? Because it's not about the movie. It's about the value and relational bonding of shared audience experiences.

Many content platforms allow viewers to watch a video simultaneously, as audience members who are not involved in the action. You can chat with other audience members to share your thoughts, feelings, and reactions. Some streaming services have this available; right before you play a movie on your TV, you can invite friends to watch it with you. Other apps and platforms offer this as well. You can experience watching the video content simultaneously, even talking with each other, despite the fact you are not in the same space.

Shared Engagement Experiences

These experiences add yet a deeper level of connectivity. In this level of experience, participants are engaged in the event beyond audience participation. For our one-year anniversary of Hearts Like His, we did a game night. Since it was January of 2021, pandemic protocols were still in place, so we streamed the event live on Zoom. I created some questions specifically about Hearts Like His, plus there were questions taken from the teaching topics throughout the year. As questions were presented on slides, participants from all across the United States raised their hands to answer. We celebrated when the answers were correct and laughed when the answers were wrong. The winners received Amazon gift cards.

The opportunities for engagement experiences vary based on your creativity and your platform. As you engage in online conversations, there are things you can do intentionally to add to the level of connectedness in shared experiences. Have participants bring items from around the house to share on a video chat. Whether in a one-on-one conversation or a group chat, have everyone wear their favorite sports team jersey, bring photos of hunting or fishing trips, share souvenirs from your favorite vacation, bring your go-to spice and share why and how you use it in your favorite cooking dish, or whatever your interest! On Christmas Eve we did a candlelight service where everyone on Zoom had a lit candle in view on their screen. When we played a lyric video to "Silent Night," everyone turned off the lights in their rooms, and the whole Zoom screen was lit with candles. It was actually pretty nice! It was a shared engagement experience. These are simple ways to initiate shared experiences via video.

Video Games

Many of today's video games have community opportunities built in. Put on a headset and play along with others around the world. Start to build relationships with people through playing video games. When I (Jeff) was in my college years (25+ years ago), we would play Nintendo64 multiplayer video games like *Goldeneye* or *Super Smash Brothers*. Four people could sit in the same room, watch the same TV, and play video games together. A couple years after the Nintendo64 era, Microsoft XBox (and Bungie's Halo) brought on a new era of video gaming. We can now play online video games.

Community is not limited to physical space anymore, according to gamers. They are playing multiplayer video games like *Destiny 2* with people they probably will never meet physically, yet they are completing major campaigns together. For video games that don't have audio chat capabilities, like *Fortnite* for example, gamers often use a separate software called Discord to create temporary community space where people can connect.

There are incredible stories of gamers praying for other gamers while playing. Remember, the anonymity of digital allows for incredibly authentic conversations. Some of the most honest conversations I've had with strangers took place while playing video games. Gamers can easily strengthen relationships with other gamers.

Remember the old video games like *The Sims*? There's a sub-genre of video games called massive multiplayer online (MMO). These video games are not so much about beating the big bad villain of the game as much as living and thriving in a video game community. Literally, gamers will play games like *Final Fantasy*, *Rust*, or *Black Desert Online* for years. Players will join tribe-like guilds and sometimes spend years building relationships with other guild members (scattered globally) in the video game. If you're a Christian in these games, what a great opportunity to B.L.E.S.S. someone in a video game!

Virtual Reality

VR is similar in experience to video games! You can play video games like *Population: One* (kind of like *Fortnite*, but in virtual reality) and engage in the community in this space. Build relationships while running around virtual worlds shooting people! B.L.E.S.S. people while playing the game. (No, we are not changing an S in B.L.E.S.S. to Shooting—although gamers have the freedom to contextualize the framework to that!)

Also similar to MMO video games, virtual reality has virtual worlds such as AltspaceVR, RecRoom, and VRChat where people spend massive amounts of time. These worlds attract different types of people and are highly relational! We won't dive into the specifics of these worlds in this book, but you can read my (Jeff's) book *VR & the Metaverse Church: How God Is Moving in This Virtual, Yet Quite Real, Reality*, in which I detail the characteristics of each world and talk about what ministry looks like specifically in the worlds. You can get more information on *VR & the Metaverse Church* at http://leadnet.org/vrbook.

Jate Earhart, a digital missionary as well digital media and design artist who lives in Missouri, shared the following story on Facebook:

STORY

I got to know a Satanist from the UK through streaming on Twitch.

We talked a lot about our beliefs, became friends, played games together, and she even stuck around for our Bible discussions (while we played Golf With Your Friends).

She attended a number of VR churches in the network of friends I've been blessed to be a part of. . . .

And today, all of us who came alongside her, loved her, and discipled her got to witness her baptism (the first in her family) in a VR world she created after months of work to house a church she's helping start, in the game where she's now leading her friends to Christ.

I honestly don't know what to say other than that if I do nothing else with the rest of my life, I will still die smiling after today. This mattered–it had purpose. Money can't buy what today was.

Shared Experiences in Apps

Digital technology (mobile phones, social media) is so driven toward shared experiences in apps that it is almost easy. Something as simple as playing *Wordle*[48] (or many other mobile phone games) can connect people, as can watching a movie/TV show together. I (Jeff) developed close friendships over long distances using the Nike+ app while I was training to run half marathons. Find common ground between you and an individual digitally, and invite them to share that experience with you. It's not just about sharing the experience, but dialoguing with the individual about the experience! Engage with people about the shared experience to build the relationship.

Shared Learning Experiences

Online learning is a big deal! College classes, certification programs, and coaching programs create endless opportunities to connect with absolute strangers from around the world online. Often, permanent digital friendships form from these temporary environments. I (John) went through a three-month online business coaching program in 2021. There were times when all 200 clients were in large Zoom meetings for instruction, but we were also assigned to smaller groups of four. A year after the program ended, the three women who were in my group, living in different countries, still chat weekly on WhatsApp and meet several times a month on Zoom.

Transformation is a big business. Fitness, health, finances, relationships, and spirituality are the four main areas where people want to see transformation. In these growth experiences people often share their stories, including their weaknesses, challenges, and failures. There are great opportunities for social listening that often provide additional opportunities to follow up with people directly about things they shared. These are great contexts that could bring you closer to someone with whom you can build a relationship and share Jesus.

Global Daily Prayer Meeting: Audio-Only

A friend of mine, Christopher Harris, executive pastor at Crossover Community Church in Tampa, Florida, leads a prayer group with eight other moderators on Clubhouse, an audio-only app. Everyone has a profile picture and a bio, but all interactions are audio-only. November 2022 will mark two years that they have been meeting every morning for about an hour, Monday through Friday at 7 a.m. EST. Their daily live digital attendance averages 350 people from all over the world: the United States, Canada, Ghana, New Zealand, and France. Yes, I said daily! There are 500 daily replays as well.

A typical meeting consists of teaching, prayer, praise reports, celebrating birthdays and anniversaries, and ending with a dad joke. And if they forget the dad joke, they are quickly reminded of its necessity. As people share their prayer requests, they inevitably share things that are very personal. Christopher and his team have created an online space where people from different cultures and backgrounds can feel safe and gain a sense of belonging, value, and support. This is a strong digital community that has been built through consistent, daily, audio-only conversations. There's no way you can tell them that this is not a church gathering. Some members of the group who tune in from Canada and the UK put the prayer call on speaker in their break rooms at work so their coworkers can listen in. One woman sends the recording to her mother in a nursing home so she can listen every day with her friends.

As I talked with Christopher about how he built a sense of community on Clubhouse he said, "Community is not just built on physical presence but the audio and how you're connecting to one another. It's not just what you say, but how you say it. When we gather for church [physically], we can actually assume that we're answering people's questions and we're actually not."

There are several things he and his team have done intentionally to foster togetherness in this global prayer community. First, the teaching answers questions people are asking. Second, they personalize the experience because when people have questions or concerns, they have the opportunity to be heard directly by the leadership. Third, every day during prayer they call out the name of each person who is participating. Overall, he said, "The questions being answered and the relatability adds value to people's lives. And when going through trials you're looking for community and looking for relief. You can't always get that from church."

So what has God been up to in this prayer group? They have documented over 1,000 miracles and answers to prayer! Yep, they have been keeping track with a spreadsheet. Wow! On one of their live prayer calls, a woman asked for prayer for her son, who was shot the night before and was in surgery during the prayer time. He made it through surgery successfully! What a blessing to have 350 people praying precisely at such a critical moment!

On another occasion, one member of the group had a daughter who had just had a baby. Due to complications, the woman's daughter ended up being on a ventilator. This time the woman asking for prayer was doing so from her daughter's hospital room at 7 a.m. As they were praying, the woman was telling them that the numbers on her daughter's machines started changing, her vitals were improving—right at that moment while the church was praying! The mother even took pictures of the hospital machines as the numbers were changing. Within several hours, her daughter was off the ventilator, sitting up, and holding her baby—all before 11 am. Where I'm from, we call that a miracle. There have been several salvations as a result.

Although this prayer group didn't know about the B.L.E.S.S. framework, notice principles of B.L.E.S.S. at work. They are praying, listening, enjoying a shared experience, serving digitally through care and support, and sharing each other's stories. It's no wonder that as a result several people have decided to follow Jesus.

What Do Shared Experiences Look Like in the Metaverse?

We need to appreciate that virtual reality is not meant to be experienced alone. Headsets are designed with built-in microphones. The games that are successful are ones that are meant to be played with others. Communities are designed so that people will talk with others. There are virtual reality worlds where people, I (Jeff) mean avatars, will sit around a campfire on benches and just talk. Other worlds include a comedy club, where you sit and listen to avatars do their routine. Bars and pubs in virtual reality are a thriving business. The point? Most virtual reality can be a shared experience.

My friend Chedders, truly a biblical person of peace in the AltspaceVR world, built a prayer garden in the platform. Through that world, he has had innumerable shared experiences with people. Here are two of many stories Chedders has to share.

STORY

I met Chap4Green there, a United Methodist from New York State who was exploring virtual reality and ended up stopping by. Not only did he enjoy the prayers for him and his wife's health, but he was inspired by how many lives we touched by offering to pray for the world-hoppers who would show up. He joined us in praying for other people, first silently and later with compassionate words. Over time, he shared with his pastor what we were doing, got him to visit on a separate headset, and even prototyped a side-by-side 3D screen in AltspaceVR in hopes of allowing VR people to connect more with pastors by seeing the pastor's physical face with depth. I happened to have a trip planned to New York, so I was honored to be able to drive five hours to visit Chap4Green (who did wear all green) and his church, and I even got to eat out with the pastor! I look forward to discovering more ways to minister in VR with Chap4Green

Christian came to A Prayer Breeze with lots of questions, having grown up a Jesus follower but finding himself caught off guard by people living out their faith in VR. At first he regularly came for prayer, but over time he became excited to pray for others. Surprisingly he even started greeting people with "Can I share the gospel with you?" a few months in, which (although courageous and in the right spirit) sometimes scared people away from receiving prayer. After he started asking whether the person wanted prayer first (then sharing the gospel after the prayer), we became a great team. He was baptized for the first time in a local physical church on Valentines Day, 2021. Now God is blessing him with a romantic relationship, and I hope that he will rely on God (not himself) to produce fruit whether he is in the VR or local church.

The beauty of the shared experience is that you continue building trust with the person. You're strengthening your relationship and showing that you understand this person, valuing who they are. Casual shared experiences like this will go far to strengthen relationships, but even shared experiences will have minimal relational impact compared to what comes next.

As you engage in these various experiences, people will open up and share more of their lives with you. During these shared experiences you are still praying and listening. Keep an ear open to needs they express. This will create opportunities for the first S in B.L.E.S.S.: Serving.

ACTION STEPS: Enjoy Shared Experiences

- Think about the various kinds of shared experiences we discussed. Do you know of birthdays, anniversaries, or other special events that you can help someone celebrate digitally?
- What do you do for fun, digitally? Do you play video games or hang out on social media or in specific groups or communities? What about posting photos on Instagram or videos on TikTok? How can you connect with others online who have similar interests as you?

» What kinds of shared experiences are you currently involved in? Start doing some social listening the next time you're scrolling through. Who can you start a conversation with through direct messages?
» Make a list of 3-5 people you feel like God has highlighted to you. Who can you invite this week to share an audience or engagement experience with you?

CHAPTER 6

To B.L.E.S.S., Serve as Christ Served Us

DAVE SAYS

Jesus told us pointedly, "The Son of Man did not come to be served, but to serve" (Matthew 20:28, NIV). He modeled for us that once you begin with prayer, listen, and share an experience with someone, there is a good chance that by then you'll have discovered how you can best serve the person God is asking you to bless.[49]

WHETHER YOU HAVE GONE SEQUENTIALLY through B, L, and E, or if you have the opportunity to jump right into a serving opportunity, this chapter is full of stories that can inspire you to serve people God has connected you with online. Remember, as you serve, you are still praying and listening, and maybe still sharing an experience together.

Lost and Found on Twitter?

In his book *MetaChurch*, Dave Adamson shares some interesting ways he uses social media. He describes how to use Twitter's geographic hashtag search to see what people who live near you are searching for online.

For example, you can search the hashtag "MovingDay" within a 10-mile radius of your church building to find and welcome new residents who have moved to the area. A couple of years ago, I was searching the hashtag "IHateMondays" when I found a tweet from a local couple who had lost their dog. Using the church account, I quoted the couple's tweet and asked our church family to contact the owners if they saw the dog. A few days later, the couple found their dog, and tweeted at our church to thank us for our help. This was a couple who had never set foot inside our church building but were now thanking our church community for helping them in their moment of need.

John's DoorDash Story

When my wife, Jeanine, and I (John) had our first two daughters, we had been members of a church and local community for about 10 years. I was an executive pastor, and Jeanine was the global missions pastor. We were surrounded by many people who were looking for ways to help, serve, and support us. We received gift baskets delivered to our door by the kids ministry, and some of our friends arranged two weeks' worth of meals. We were surrounded by a physical community as people brought meals to our house everyday.

But when my wife gave birth to our third daughter, we had moved to a city an hour away and had only been there for about five months. We did not have that physical community, as we were new to the area. We were definitely feeling the difference of experiencing this joyful transition away from our community. So when a friend who now lived an hour away sent dinner via DoorDash, it meant the world to us because they did not let physical distance stop them from serving our family. They thought outside the box about what they could still do to be a blessing to us. And were they ever! This woke me up to the opportunities we have today to serve people through digital means.

In this day and age, so much can be accomplished online or through apps on our phones. The sky's the limit when it comes to how you can meet the needs of people through digital space regardless of geographic proximity. Nowadays you can order a mobile car detailing service, send meals, have someone's groceries delivered, and so much more! As people discuss their lives online and the challenges they are currently facing, keep listening for ways you can serve by meeting their needs. Let's unpack the significance of Serving, the first S in Dave's B.L.E.S.S. framework.

As Jesus went around preaching the good news of the Kingdom of God, his preaching was accompanied by meeting people's tangible needs. Serving opens up the heart to new relationships. When we serve with the genuine desire to meet someone's need, they can feel that we have put their needs before our own. Serving requires some level of sacrifice that many people recognize as an act of love (whether they articulate it as such or not). The one being served feels seen, heard, valued, and honored when we serve.

"I Am Among You as One Who Serves"

While the environment in which we serve does not matter, what *does* matter is our posture, our attitude. As digital missionaries, we began with prayer, inviting the Holy Spirit to guide us. We followed the Spirit's leading to begin (or renew) a relationship with someone. That relationship began by our listening to the individual. We strengthened that relationship by inviting them into shared experiences digitally or in the metaverse. All of this was done out of humility, giving the other person's needs a higher priority than our own. As we enter into serving our friends, we maintain the posture of humility.

> Serving opens up the heart to new relationships.

Jesus' ministry was marked as much with serving as it was with his teachings and miracles. In fact, some of his miracles were all about supernaturally serving people. He served people when he turned water to wine in Cana (John 2:1-11). He fed the 5,000 (Matthew 14:13-21) and the 4,000 (Matthew 15:29-39). Jesus modeled humility in service to others. He often needed to teach and remind his disciples of this.

On one occasion, after reminding his disciples that he would be betrayed and go to the cross, the first thing on their minds was who was the greatest. "But not so with you. Rather, let the greatest among you become as the youngest, and the leader as one who serves. For who is the greater, one who reclines at table or one who serves? Is it not the one who reclines at table? But I am among you as the one who serves" (Luke 22:26-27, ESV).

While his disciples were trying to pursue higher positions, Jesus used the opportunity to teach the Kingdom value of humble servanthood. On another occasion, after sharing the Last Supper the night before going to the cross, Jesus washed his disciples' feet. When he was finished he said, "For I have given you an example, that you also should do just as I have done to you" (John 13:15, ESV). Greatness in the kingdom of God is the opposite of greatness according to the world's values. Dr. Martin Luther King, Jr. summed it when he said, "Anyone can be great because anyone can serve."

> "Anyone can be great because anyone can serve."
> –Dr. Martin Luther King, Jr.

If we're honest, it's easy to lose sight of why we're doing things. Jesus' reminder to his disciples is also a good reminder for us as we serve.

How Can We Serve Digitally?

When someone expresses a need online, it's easy to start with a framework of "Well, there's not much I can do since I'm not in person." This limits your creativity and problem-solving ability. Instead of thinking about what you *cannot* do, think about what you *can* do. In this chapter, we want to share some ideas that are intended to get your creative juices flowing.

Digitally Serving Through Written Prayer

One benefit of written communication is that the person can go back and read your words as often as they want or need to. For example, that's the power of typing a prayer for someone instead of just putting the prayer hands emoji. The heartfelt words of a typed prayer can be something that the receiver can revisit when they need to be encouraged. Live prayers online, which many people do, are also stored in online platforms and have the same affect. The prayer does not have to be long to be significant. When someone hears/reads you talking to God on their behalf, it is very meaningful.

In June of 2022 Megan Carter, from Church Anywhere, received a message on TikTok from Lisa. (We decided to share it in its authentic form. People do not care about grammar when they are asking for help, and neither should we.)

STORY

Lisa wrote Megan and said:

Hey, you prayed for my sister before she has Rett Syndrome. Can you please pray for her, my dad and twin? They all have covid. The illness makes it dangerous for my sister and my dad went to the er today. Also say a prayer for my husbands family they lost his grandma. I'm sorry to bother you, but you are the one person I know that will pray with and for me.

Megan told us:

When this message came in I grabbed my phone and instantly responded with a written out prayer for her. I also let her know that she was not bothering me and I'm here for her anytime. To which I received this response, "Thank you so much Megan! The one thing this app [TikTok] did for me was show you and you have been an inspiration and motivator for me. Thank you so much for praying for my family." It took five minutes out of my day to stop and pray, but it's not only that. She had watched me through my videos on TikTok and had seen how authentic and genuine I am online, so much so that it built trust enough for her to reach out to me and know that I would be there for her. It's all about relationships.

I don't want you to miss the power of what Megan did here. There is an often-overlooked ministry opportunity for us in moments like this. So let me take some time to unpack the power of written prayers.

God Uses Prayers, Not Emojis

It's now a trend when someone has a prayer request to type "praying" or to send the emoji with prayer hands. It's trendy, socially acceptable, and assuring on some level for the requester to see responses to their request. However, let me mention a few concerns. First, when we say "praying" we typically mean "hoping." It's kind of like saying, "I'm praying for you," which sounds like a more spiritual way of saying, "I'm hoping things get better." Let's not tell people that we are praying if we are not actually praying. Second, real prayer shifts things in the spiritual world. Real prayer can push back forces of darkness. Real prayer can move Heaven on someone's behalf. Prayer emojis do not do that.

When you see an online prayer request, I want to encourage you to take a moment and type a prayer. It might not be all of the time but certainly more than what you may currently be used to. It does not have to be long, but it will be more meaningful to the one requesting prayer. It will stand out among all the other emojis and "praying" comments. A written prayer is more meaningful for several reasons.

First, it lets the person know that you took their request seriously and actually prayed. A written prayer communicates that you value the person enough to take time to think about what you want to say to God on their behalf and actually pray it. They know that you stopped what you were doing to spend time on their request.

> Real prayer can push back forces of darkness. Real prayer can move Heaven on someone's behalf. Prayer emojis do not do that.

Second, it shows that you were, like Jesus, "moved with compassion" (Mark 1:41; Luke 7:13) to act upon hearing/reading their request. It shows that you did not just type an emoji and keep scrolling, but you stopped and acted thoughtfully and compassionately. Regardless of a person's level of faith, it is a powerful thing to hear/read you praying over them and their situation. When things get bad enough, even atheists will surprise you by asking for prayer.

Next, when you pray for someone, it ministers to their heart and can be very encouraging. The heartfelt words of a typed prayer are something the person can go back and read as often as they want to and need to. You'd be surprised by how many prayers get screenshotted so people can have easy access to them for future reference. People who do not know how to pray can and will use your words as their own. Think about how we use the prayers written in Scripture, especially during times when we do not know what to say. People will also use your prayer as a reminder to be aware of God in times they forget about him.

Also, when you type the prayer in a public feed instead of a direct message, it communicates a bold faith by being vulnerable in a public forum. It's an encouragement to other Christians to go beyond the emoji. It models for other Christians the simplicity, power, and effectiveness of written prayers. I used to be the prayer emoji and "praying" guy until I saw someone type a prayer in response to someone's request. I thought, "Wow, I can and should do that too!" It really ministered to me, and it was not even my prayer request. You'll find that your prayer post will get a lot of reactions as others agree.

Finally, it's a witness to unbelievers who will see your love and care in the words of your prayer. Written prayers are another way to show the compassion and love of true Christianity.

Digitally Serving by Sending Love

Every year I would send flowers to my mother for Mother's Day and her birthday. I would go online, and although I was in California and she lived in Arkansas, it was a simple but meaningful process. She would always send me pictures of the flowers and gifts I would send with a big smile on her face. I bring this up because even though I had done that for years, for some reason I did not consider other ways, like my DoorDash story, of how we could serve people through digital means. There may be people in your circle of online influence who have a current need you are aware of. Maybe you have not taken time to sit and think about how you could serve them from a digital distance. Although this book is mainly about reaching out to those who are not saved, these principles help to build any kind of relationship. For some of you, it may be easier to practice these steps on people you know before using them on people you are less comfortable or familiar with.

Digitally Serving by Resourcing

Not all serving through digital space requires money. Sometimes serving means connecting people to resources they need. Even if you do not provide the actual service, the role you can play in connecting them to others can be a tremendous service. A great example of this is what Anne Bosarge of The Chapel Online does on a regular basis. The Chapel Online is an online church on Facebook. Although Anne lives in Georgia, many of the over 10,000 participants in her ministry live in Africa. There are often needs that arise that (because of her extensive network in Africa) she is able to meet because of significant connections that serve people on another continent. Here are a couple stories of how Anne has seen people serve digitally.

STORY

There is a young man named Frank who has five younger brothers and sisters–they are all orphans. Frank and I were connected for quite a while through The Chapel Online. We worked through Compassion International to get Frank's brothers and sisters sponsored and into school. Then I introduced Frank to another young man, Joshua, who lived about two hours away from him. Joshua was able to visit Frank and help us get Frank registered in a tech school there, and someone from the United States is paying his tuition to become an electrician. Joshua visits from time to time and checks in with Frank. About a month ago, one of the younger kids needed some medical attention. So we connected Frank with a doctor two hours away who was able to take care of the problem for us and perform the tests necessary to determine the right course of action.

STORY

Joshua, from the story above, leads an organization called Lift Up Girls. The group empowers young women and girls (as young as 13, who are often mothers) to earn a living wage and provides leadership development for them as well. Lift Up Girls has 12 community centers in and around Jinja. Well, Derrick (a young seminarian), John (another friend I connected with), and Joshua didn't know each other but were all doing ministry in the same area. So I introduced them, and now all three are partnering to meet the needs of the women of Jinja, and they are close brothers in Christ!

What Happens When We Serve

You have probably heard the popular colloquial phrase, "People don't care how much you know until they know how much you care." The truth of this is seen in how the B.L.E.S.S. framework works. When we serve, it softens people's hearts toward us. When we serve, it gives our message credibility. When we serve, it builds trust and connection. When we serve, it unlocks hearts in a way that not much else does. Think about how some people try to apologize by doing something kind for the person they offended. It's an effort to unlock forgiveness in the heart of the offended so the relationship can be repaired. In the early stages of dating, traditionally the man tries to win the lady's heart by serving: opening doors, buying dinner, etc. Typically, the relationship advances to the point where each one is trying to learn what the other needs and wants, and each makes an effort to meet those needs to show love and honor. When we serve people, over time, people may ask why you are doing what you are doing, why you care so much about them. This is a clear open door to share the gospel. However, sometimes that question is not asked, and it is up to the missionary to discern the right time to shift a normal conversation into a spiritual one.

Is It OK for Serving to Move the Relationship Offline?

The key to being a digital missionary is that we desire to see lives changed physically as well as digitally. We want our relationships to not just cover our digital or metaverse selves but to also impact the physical lives of the people we're building relationships with. This is why the "online to offline" standard we mentioned earlier is so important and so relevant here as we're focusing on serving. We want to see the fruits of the Spirit in physical as well as in digital/metaverse spaces.

As digital missionaries, we have the opportunity to serve people digitally, physically, or a combination of both. We don't need to draw a line and say that serving has to be digital or that it must be physical. Ultimately, find the environment that connects best with the person you're developing a relationship with, and serve them. If this is the point at which your digital relationship moves to physical, so be it. God moves in the physical space as well as he does digitally (and vice versa).

We want to see the fruits of the Spirit in physical as well as in digital/metaverse spaces.

Serving in the Metaverse Kinda Feels Like Digital

Just because you're engaging with someone in the metaverse does not mean you're limited to serving that someone in virtual reality. Let's think about it. You've connected with someone in VR and prayed about talking with them. You feel the Spirit leading, and you start talking by listening. You want to build a relationship, so you invite your friend to play *Among Us VR* or to explore some worlds in RecRoom. Maybe God can lead you to think of ways to serve this person in virtual reality, but if you're thinking, "Hey, maybe I can mail them a book or give them an UberEats gift card," these solutions are just as valid. In fact, you'll probably gain more credibility by engaging with them in digital or physical space.

The sticky part here is that you don't want to be accused of being a "creeper," as my (Jeff's) kids call them. Creepers are people who try to gather information online that the person is not ready to give up. So a 40-year-old man in virtual reality trying to get a 22-year-old woman's email address? This is a creeper. Don't be a creeper. Use personal judgment when appropriate. But don't let missional imagination be hamstrung because you feel boxed into one environment.

Be creative! Missional imagination may start to open up innovative opportunities that have yet to be imagined! Short of that, it should be relatively easy to move a metaverse relationship toward digital at this point, or even all the way to physical, if that's possible. How do we discover what our people need? Check out Dave's R-P-M-S!

DAVE SAYS

The B.L.E.S.S. practices are intentionally sequenced in an order that builds a bond of friendship and helps people feel loved. It's what Jesus did, and it's hard to improve on that. It is in the context of friendship and love that others will often confide in you how you can serve them. For example, I have a friend Michael, who told me about a tragic event from 20 years prior that haunted him every day. He was looking for forgiveness from the tragedy he had caused. I was able to serve him by telling him how he could be forgiven and how God redeems our pain.

Other times, people are not as forward with what they need and how you can serve them. That is where this simple tool of R-P-M-S can help you discern how to serve someone.

Just as RPMs (revolutions per minute) can help you gauge the health of an engine or motor, I have used a similar acrostic as a way to monitor my own personal wellbeing for many years. It comes from Luke 2:52 (NIV), which describes Jesus' own personal development: "Jesus grew in wisdom and stature, and in favor with God and man." He grew in these four areas:

» **Relational:** He grew in favor with men.
» **Physical:** He grew in stature.
» **Mental:** He grew in wisdom by developing his mental capacity.
» **Spiritual:** He grew in favor with God.

Every morning I write these four letters (R-P-M-S) at the top of my journal and give myself a score from 1 to 10 on how I'm doing on each of them. I call it a "two-minute checkup." It has been a great help to me personally. More recently, I've used R-P-M-S to find out how the neighbors and friends I want to bless are doing. Let me briefly give you a few questions to ask about the neighbors you love.

How's my neighbor doing **RELATIONALLY**? How is their home life? How is their marriage, dating, or family life going? Do they have close friends? How are their relationships at work? Do they have a healthy circle of friends?

How's my neighbor doing **PHYSICALLY**? How is their overall health? Does their energy level seem good? Are they getting regular exercise? Are their eating habits healthy? Do they mention not being able to sleep?

How's my neighbor doing **MENTALLY**? Are there any signs of anxiety? Is there any indication of depression? Have I noticed any mood swings? Are they learning? Are there any unhealthy thought patterns showing up?

How's my neighbor doing **SPIRITUALLY**? Do they sense something is missing in their life? Are they willing to have you pray for them? Do they display a spiritual curiosity? Do they initiate spiritual conversations? Are they moving closer to God?

Just as we serve ourselves and God when we monitor our personal R-P-M-S, we can bless others in these same four areas. Once you identify their needs, serve them![50]

Serving Beyond Your Wildest Imagination

Henry Blackaby, through *Experiencing God* and other works, talks about how God speaks through the Holy Spirit, through his Word, and through others.[51] I (Jeff) love how God speaks into our imagination, guiding us as we seek to connect with someone. I've also seen incredible ways other missional imagination flows through others and how God uses imaginative, innovative community to collaborate on Kingdom ideas.

Curious about what serving in digital or the metaverse can look like? Need some help imagining digital or metaverse ideas? Jump into

Digital Church Network's FAM, at http://fam.digitalchurch.network. Our online community exists in part to help digital missionaries like you do the ministry that God has called you to do. So if you're looking to jumpstart your missional imagination or looking to celebrate with others what God is doing through you as a digital missionary, create an account for free and engage in a community that is designed to help your digital/metaverse ministry.

ACTION STEPS: Serve

- Spend some time this week and apply the R-P-M-S to one of your neighbors. Then send a typed or voiced prayer.
- Take 10-15 minutes and join your favorite social media or metaverse community. Ask the Lord to highlight who you can serve. Do some social listening, and prayerfully consider how to serve the person or family in one of the ways we discussed in this chapter.
- The next time you see someone digitally ask for prayer, type a prayer, hit "Send," and trust that God will be at work.

CHAPTER 7

To B.L.E.S.S., Share Your Jesus Story

DAVE SAYS

My hunch is that you could quickly get on board with the first four B.L.E.S.S. practices we discussed. Yes, they require some intention and effort, but they aren't particularly difficult to carry out. For the most part, they don't push you too far out of your comfort zone. However, this final S makes many well-intentioned Christians nervous or reluctant to move forward. But you can't spell "B.L.E.S.S." with just one S. And you won't likely help anyone discover the love of God without using your words—and your story.[52]

Why Your Story with Jesus Matters

IN THIS CHAPTER I (JOHN) want to share some biblical insights and practical tips on why your Jesus story matters and practical tips on how to do it. One of the things that accelerated and expanded the influence of Jesus' ministry more than anything else was that people were telling stories of their encounters with him. John 4 records the longest conversation that Jesus had with any individual, and he had it with a Samaritan woman. Because of her powerful encounter with Jesus at the well, she left her water pot and ran back to her city to tell everyone her story. John 4:39-42 (ESV) says that as a result:

> Many Samaritans from that town believed in him because of the woman's testimony, "He told me all that I ever did." So when the Samaritans came to him, they asked him to stay with them, and he stayed there two days. And many more believed because of his word. They said to the woman, "It is no longer because of what you said that we believe, for we have heard for ourselves, and we know that this is indeed the Savior of the world."

On another occasion, in Mark 5, Jesus cast thousands of demons out of a man who had been isolated and left to live among the tombs because he was so violent and dangerous. When the man was freed from demonic oppression, he wanted to follow Jesus and the disciples. As Jesus and the disciples were getting in a ship preparing to sail across the Sea of Gennesaret, the man attempted to get into their boat so he could follow Jesus. Jesus refused to let the man join them.

Instead, Jesus said, in Mark 5:19-20 (ESV), "'Go home to your friends and tell them how much the Lord has done for you, and how he has had mercy on you.' And he went away and began to proclaim in the Decapolis how much Jesus had done for him, and everyone marveled." When Jesus cast the demons out of this man, they went into a herd of 2,000 pigs, causing the pigs to run off a cliff into the sea and drown. The residents told Jesus to leave because they no longer wanted him around because he literally cost someone a personal recession.

Even though Jesus was forced out, he later came back to that area. When people heard that he was returning, the Bible says they brought those who were demon-possessed, those who were sick, those who were unclean, and those who needed help, hope, and healing. They packed out the place when they heard he was coming back.

How did they know he could do something about their situation? Because when Jesus wanted to start a ten-city (Decapolis) revival, he started with a demon-possessed guy. The man obeyed what Jesus said, and he told everyone what the Lord did for him. There's no doubt that just seeing him sane was a testimony to all who knew about "the guy who lives in the tombs."

In both of these examples, neither the woman at the well nor the man who was delivered from demons quoted a list of verses from the Old Testament as they shared their story. They were so excited about their transformation that they did not worry about how to say what they wanted to say, who would reject them, or what would happen if their story did not come out right. They just shared what happened to them and who made it happen. Allow me to share one more biblical example before we move on, one with a unique factor that you may run into, one with an important takeaway.

"I Was Blind, and Now I See!"

John 9 tells the story of Jesus healing a blind man by spitting in the dirt, making clay, and putting it on the man's eyes. Jesus told him to go to the Pool of Siloam and wash his eyes. When the man did that, he was able to see. This would seem to be good news for *everyone* to hear! However, since it took place on the Sabbath day, the Pharisees disapproved and accused Jesus of being a sinner. When they asked the healed man about his opinion on the matter, he answered, "Whether he is a sinner I do not know. One thing I do know, that though I was blind, now I see." (John 9:25, ESV). Ultimately, this guy was saying, "I'm not interested in your religious argument. All I know is I have been changed. Something happened to me! I now have a before and after. I was blind, and now I see!" This is such a great response: "I don't know about religion, but I know about results!"

This is great for two big reasons. First, you don't have to be an expert in the Bible to share Jesus. You can just share what you know and share your experience. Second, people are more interested in your story than a bunch of Bible verses. Those can come later, but first they want to hear about the change in your life. Share what you know, and leave the rest up to the work of the Holy Spirit.

Don't Call Your Pastor!

Don't call your pastor to do what God positioned you to do. Your pastor has a calling, and so do you. Your pastor's primary role is to equip you for the work of the ministry (Ephesians 4:12), to prepare you for these moments where you can share the story of your journey with Jesus. These conversations are *your* moments. God set *you* up with the conversation and the relationship. You've invested in this person, and if you've reached the point of sharing your story (whether it has only been 30 minutes with the person or three years), this is not the time to outsource this moment to your church or pastor. Fight the temptation to think, "I wish my pastor was here" or "I wish they lived closer so they could come to my church." Listen! This is *your* moment, *your* God moment! Own it! Lean into it! Ask God what he wants you to do with it! God wants this person to hear what he has done in *your* life, not your pastor's life. After seeing the statistics mentioned earlier, you can probably understand why I say that people are more interested in hearing about your experience *with* God than hearing your pastor's sermon *about* God. You have the relationship that you have built. You have credibility that they trust. And because you are not a pastor and they are not a pastor, your story will be more relatable. When a pastor talks about a relationship with God, it's easy for it to be dismissed as something that the average person cannot experience because of the belief that the pastor has a "special connection." When you as an everyday believer tell your story, it is more meaningful to the average person.

How to Share Your Story

Although there are other topics that can come up in a spiritual conversation that you might not be prepared for, this is not one of them. This is the part that you *can* prepare for—because you are an expert on your life and what God has done in you and for you. It can be overwhelming to try to sift through your life with God on the spot if someone asks you about it (especially if you have never done so before). Have you ever thought

> When you have more confidence, it will lead to you sharing your story.

about how you would share the story of how you were saved? Where and when did God save you? What did he save you from? What are the things in your past that you no longer do anymore? How would you communicate that to other people?

The following steps are a place to start if you really have no idea where to begin. My hope and goal is that you will be prepared for this part of the conversation so you can have more confidence. When you have more confidence, it will lead to you sharing your story, and then to doing so more often and with less fear. The Holy Spirit will surprise you with what he does in the lives of other people who hear what he has done in your life. That alone will be a motivator for you to share again and again.

What's Your Salvation Story?

I shy away from trying to give a formula for how to share your story. Don't think of these questions as a formula, because they're not. In fact, when you share your story with someone, you may end up sharing events in a different order based on the other person's questions and interests. So this is not a step-by-step directive but a list of questions that people often want to know the answers to as they relate to your experience with God. Think through the major points of your salvation story and what your life with God has been like since you were saved. What was it like when you were saved? Some people can identify one significant moment or turning point. For others, there was a series of events that moved them closer to their decision. Even still, for others, they do not remember any particular experiences or events. They just know that at some point they trusted in Christ. No matter how God moved in your life, own your story. Own the unique way God led you to faith in Christ, and just share your journey.

How to Craft Your Story

Consider the following questions as you begin to draft the story of your life before and after Jesus. This is just a starting point, and it certainly doesn't need to be perfect—just honest.

1. **What was your life like before meeting Jesus?**
   » What were your values and priorities?
   » What was/was not important to you?
   » What was your personality like?
   » What was the quality of your relationships?
   » Was there anything notable about your emotional health? Any periods of hopelessness? isolation? pride? emptiness?

(If this section is difficult to write first, come back to it after you talk about how Jesus changed your life.)

2. **How did Jesus change your life?**

Include the circumstances that led to your decision. Some of them could be the significant life events we mention in Chapter 6.

This is also where you can include the gospel without sounding religious. When you talk about the gospel as part of your experience, it does not sound as though you are telling someone else what to believe. For example:

» **Without the gospel:** "I'm a recovering alcoholic. Jesus set me free, and now I don't drink anymore."

» **With the gospel:** "I'm a recovering alcoholic, and Jesus set me free. When I learned that in him I could be a new person because he died on the cross for me, when I really grasped that my sins could be forgiven by this God who loves me, it radically changed me. I tried this program and that program but wasn't able to really break this. But when I had this encounter, I learned that God's Son died in my place for my sins. That's what impacted me! And because of what I received from God, I don't drink anymore. I was getting drunk to fill a particular void that his love now fills. And now I'm so different."

Obviously the "with the gospel" version was much longer, but please don't focus on that. (I added extra words to make it sound more natural and conversational.) What I want to show you is how you can naturally include the gospel in your story without sounding religious or sounding like you are trying to force your beliefs on someone. This approach is just you sharing your experience, and it is usually better received when someone really wants to listen.

Go into detail about how you are now different from what you were like "before Christ." How has your attitude about your life, your purpose, or your worth changed? How does it feel to know Christ lives in you? Talk about the peace and joy you now have. What's it like to be certain about where you are going after this life is over?

This can also be a good time to address myths about the Christian life that the other person might believe. For example, some people believe that if they become a Christian, they are expected to live a perfect life. You can share that being a Christian is not about being perfect and without mistakes, but it is about being forgiven in the eyes of God because of our faith in Christ. There's freedom in living for Christ because of his grace. We have been given grace for our sins and mistakes. We have the Holy Spirit guiding us, strengthening us, helping us become like Jesus.

3. Has anything like that ever happened to you?

This question transitions from your story to their response. It promotes introspection as well as a bit of comparison. As you share, the listener is probably already identifying with your mistakes, dark days, negative experiences and emotions—but they will probably not be able to identify with the hope and joy you now have. Asking, "Has anything like this ever happened to you?" or some version of this question promotes reflection and prompts them to invite you into their thinking process.

As you think about these questions, try to put the answers together in a way that takes three to four minutes to share. This is what you want to practice sharing. As we mentioned before, the goal is not for you to have a polished presentation but to take some time to think about what parts of your story you want to share. What parts of your story would help others see how real Jesus is and the impact he can have on people's lives? Another goal is to build your confidence by thinking this through ahead of time to lessen the anxiety and overwhelm that can come by sharing it impromptu. As a matter of fact, as you tell your story to people and answer questions, you may change what you share. You will also learn to adapt what you share to what might be relevant and of particular interest to each unique person you talk to.

Great Expectations (Or No Strings Attached?)

Since sharing your story is personal and vulnerable, it's easy to have certain expectations for how people respond to you. Because this is such an investment of yourself, it is natural to have expectations of a return on your investment. Just expect to share, and trust the response and the results to the work of the Holy Spirit.

Also, you should not expect everyone you share your story with to immediately respond by asking, "What must I do to be saved?" Be free from the pressure of having to convince someone of the gospel or that your story is true. It's the Spirit's job to convict of sin. It's the Spirit's job to open hearts and eyes. You cannot do the Spirit's job. If you try to be convincing and persuasive, you can add or take away from your story, which shows lack of trust in God's work in their hearts. Be free from expecting that you have to "close the deal" in every conversation. Be free from expecting people to fall on their knees before God repenting of their sins after hearing your story. Your Jesus story is a seed of faith. It becomes part of their story when you share it. Just share your story!

As you share with different people, there will be different responses. Some people may have questions immediately and want to engage based on points that piqued their interest. Others will want to ponder and come back later with questions. Some people might just respond with, "Eh, that's good for you." Sometimes there will be no evidence that your story had an impact at all, but the Holy Spirit will use it as a reference as he speaks to the person in the weeks, months, and even years to come. There will be some who reject your story or at least your interpretation of it. They will want to attribute the positive things to something other than God. There are so many potential scenarios for responses. Keep these in mind so you can manage your expectations. Everyone is in a process, and sharing your story is only one part of it.

When believers talk about sharing their story, a few concerns usually come up. Let me (John) address those before we dive into practical tips.

Myths About Sharing Your Story

There are a couple of major challenges to sharing your story for some Christians. First, those who do not have a dramatic Saul-to-Paul story can have a tendency to believe that their story is not exciting enough to share. I struggled with this for the longest time. I heard so many testimonies from others about believers who were abandoned by their parents at a young age, overcame drug addiction, lived a promiscuous lifestyle, or other "darkness to light" stories. My story was not like that. I had a less-dramatic life story. My parents were solid believers and were married for almost 40 years until my father, a pastor, passed away. There was no alcohol or abuse in my home. For the longest time, I thought that no one would be interested in hearing my story because it was not exciting enough.

But that's what makes my story great. Because of the Christian environment I grew up in, I avoided all of that trauma and drama. However, I was still a sinner. God does not only save us from dramatic sin, but also from subtle sin. Regardless of the details, all of God's children have passed from darkness to light and have been made right in God's sight by our faith in Christ (Romans 3:22). My storytelling emphasizes the drama of the cross and what Jesus did to make me a son of God. My unique spin is that no one can consider themselves "good" compared to the holiness of God and his standard for righteousness. Sins are not just in our behaviors but also in our thoughts. Even though by many standards the world would say I have lived a good life, "all have sinned and fall short of the glory of God" (Romans 3:23, ESV).

> God does not only save us from dramatic sin, but also from subtle sin.

Second, those who were saved at a young age do not have a long period of "before Christ" time. I accepted Christ at a young age. This can create the same feelings as mentioned above, but the same principles apply here as well.

DAVE SAYS

All great stories are the same. Storytelling is an art. Every great story has characters, a setting, a plot, a conflict, and a resolution. The character is who the story is about. The setting is the place where it happens. The plot is the beginning, middle, and end of the story. Conflict is the problem in the story, and the resolution is how the problem gets worked out. Without these elements, our stories can become rambling, disconnected, and uninteresting. By simply telling your story in the three parts we suggest—who you were before you met Jesus, how you met Jesus, and what your life has been like since you met Jesus—you will include all five components of a great story and share yours in a way that is clear, concise, and compelling.[53]

ACTION STEPS: Share Your Jesus Story

Craft Your Story

Get something to write or type with, and craft your Jesus story. Use this simple framework:

- » What was my life like before Jesus?
  - Thought and behavior patterns
  - Emotional or mental stress or strain
- » How did Jesus change my life?
  - Did you have a specific moment, series of encounters, or gradual change over time?
- » How has your life been different?
  - What new life in Christ is like

Practice Your Story

Think about a friend or family member with whom you can practice telling your story.

- » Get their feedback. They might say, “Oh, don’t forget when this or that happened! Make sure you tell people that part!”
- » See if you can keep it to 2-3 minutes.

Be Ready with Your Story

Keep a digital copy of your story on your phone for easy access when you need it. Use it as a reference when sharing online.

CHAPTER 8

Moving B.L.E.S.S. Forward: Where Do You Go from Here?

What Seth Godin Taught Me About Digital Missions

LET'S TALK ABOUT SETH GODIN. If you don't know of Seth Godin, he is one of the top thought leaders in marketing/communications globally. On a recent episode of the *Akimbo* Podcast,[54] Seth was asked a question (by someone from Romania) that hit surprisingly close to home:

> What steps would you follow to build an audience that you go deep with? What is the right medium to expose yourself? [On what basis] would it be OK to do it, so that you also build momentum and at the same time take advantage of the media algorithms that are affecting our work?

That being said, I (Jeff) think Seth's answer may surprise some of you. It certainly surprised me.

> What's the smallest viable audience, the smallest group of people that you could connect with, that you could inspire, that you could change, that you could make a living with, and that would sustain you on your journey? Because the answer is not a billion people. It might not even be 100 million people or 10 million people or a million people. And the

place we begin getting trapped is saying, "It's so good. It's for everyone." How do I get the word out? The alternative is to say,"It's for someone." How can I be specific?

Modern culture has broken us all down into smaller sections. This is what Seth Godin understands, better than most, and it's what makes the B.L.E.S.S. practices so effective for digital missionaries. If you want to impact the world, you don't try to impact a billion people. If you want to impact the world, you impact an individual, and then do it again and again and again. And you help *your* individuals find *other* individuals to impact, and so on. In 2023, we're no longer dealing with one giant world. We're dealing with many, many smaller groups of people within our world, and we must learn to reach people in the sub-genres of their lifestyles.

Digital missionaries, this is your challenge—not to reach a billion people with the most influentially generic social media post on the planet, but to build a relationship with someone. Through that relationship, tell that someone about Jesus. Help them to hear that message, own that message, and then encourage them to tell that message to someone else.

Helping Others B.L.E.S.S.

Without realizing it, you've been learning a digital missionary framework that is approved by one of the top marketing thought leaders on the planet! Through the B.L.E.S.S. steps, we are leaning upon the foundation of the Holy Spirit and listening to a person in order to better understand them and to "craft" a specific message for them. Following Seth's strategy, B.L.E.S.S. communicates the message so that someone not only hears the message when we share our story in Jesus, but that person also understands that message and what it means for them. What we may need to emphasize, though, is that for the message to spread, we need to tell someone else. There is a multiplicative piece to this that we often miss. Remember, even in Seth Godin's model, the goal is to get your message (in our case, the gospel) heard around the world. As digital missionaries, essentially, it's our job to create more digital missionaries. How do we help others adopt the B.L.E.S.S. practices? What does accountability look like moving forward?

This is why an individual digital missionary "on mission" or missionary movements may be more effective for evangelism than the typical US church. The relational power that digital provides, plus the posture of a digital missionary to reach "someone" as opposed to "everyone," often makes individuals more effective than an institutional church (specifically in digital space).

From a non-biblical perspective, Seth Godin agrees. Want to make big changes? Think multiples of small. By the way, the apostle Paul said the same thing a couple thousand years ago. In 2 Timothy 2:2 (NIV), Paul says, "And the things you have heard me say in the presence of many witnesses entrust to reliable people who will also be qualified to teach others." Paul is encouraging Timothy to disciple others, just as Paul discipled him, so the gospel would propagate. Multiple teachers teaching multiples of others. One message, Christ crucified, presented and contextualized in multiple ways. Digital missionary, the world needs your voice, your imagination, and your relationships to point people to Jesus.

Remember the second half of the Great Commission? It says, "and make disciples of all nations." One of the awesome benefits of using the B.L.E.S.S. principles is that you have a simple framework you can teach others to use. That is how the B.L.E.S.S. effect multiplies and how the Kingdom of God is advanced online and in person.

Overcoming the Digital Missionary Misconceptions

Remember that list of misconceptions from Chapter 2? Hopefully now you're seeing that (as a digital missionary) it's easy to overcome these obstacles.

As a digital missionary, you see that you don't need to be technically savvy to share Jesus. Yes, sometimes people are not always genuine in digital space, but the majority of them are. There is potential in using these digital channels and communities missionally.

As a digital missionary, you recognize that, in fact, there are people in your circle of influence who need Jesus—whether they are people you are currently connected with, friends from other seasons of life, or even people you've never met before. Digital/metaverse missionaries recognize that there are people in need of Jesus everywhere, and digital is a great door to help you connect/reconnect with them.

Accountability for a Digital Missionary?

Digital missionaries like you believe you can use these tools to connect with people digitally. They take advantage of simple frameworks like B.L.E.S.S. to hold themselves accountable for sharing Jesus online. They even consider inviting others into the B.L.E.S.S. process for more practical accountability or even to mentor/apprentice someone else to become a digital missionary as well.

Looking for people to help you on your digital missionary journey? Wanting to connect with people who can encourage you on this journey? Swing over to Digital Church Network's Online Community. You'll find hundreds of digital and metaverse missionaries, planters, pastors, and practitioners. In fact, there's an online course on using B.L.E.S.S. specifically to be a digital missionary. This is a great opportunity for you to unpack these ideas with others in online community and walk with others along the path. For more information on DCN's FAM and the Digital Missionary Online Course, swing over to http://digitalchurch.network/missionary.

What's in Your Hand?

In the Old Testament, Moses led God's people out of Egypt. There was a moment when they were trapped at the edge of the Red Sea with Pharaoh's army approaching, intending to capture them and bring them back to Egypt as slaves. As the people complained to Moses, he asked God what to do. "The Lord said to him, 'What is that in your hand?' 'A staff,' he replied" (Exodus 4:2, ESV). God told Moses to use what was in his hand to bring about the miracle of the parting of the Red Sea. When it comes to digital evangelism, we are encouraging you to literally use what is in your hand. Whether it is a smartphone, laptop, or VR goggles, use what is in your hand. Do not worry about what you do not have or cannot do. God will use what you have and change the world with it.

Mom's Digital Encouragement Ministry

My (John's) mother was such an encourager. As I was growing up, she was a constant voice repeating, "Never give up! Go after your dreams! Trust God! You can do it! Don't let discouragement get you down!"

During most of her adult life, she was very sick. Most people would not be able to tell because of the genuine smile she often wore when she greeted people. She was always so excited to see her church family—the smiles, laughs, and hugs were abundant! For years, she constantly battled the pain of multiple chronic diseases. Her social activities were limited to going to church and doctor's appointments. But when she saw you, she would always genuinely ask how you were doing. I would often imagine how much more she would be able to help others if she was physically able. She loved giving advice and encouragement. She would send daily texts of encouragement to members of our family. If she missed a day, she would double up the next day.

The texts were uplifting Bible verses, poems, quotes, and personal notes. As my family and I approach the one-year mark of when she went to be with Jesus, I cannot help but think about the biggest surprise I learned after she passed.

As we called people to let them know she had passed, a common response was, "I knew something was wrong because I didn't get a text for a few days." Several people tried to call, but the calls went unanswered. It was evident from the reactions of church members, physicians, military friends, etc. that my mother had been texting a *lot* of people.

Here I was wishing my mother was healthier so that she could get out and serve the way I thought she wanted to, but she had created her own digital encouragement ministry! She used what was in her hand to encourage people across the country. She would not have called it a ministry, but that is exactly what it was. So many people talked about how important it was to them to receive her texts. They talked about how those texts often made their day, reminded them of God's presence, or simply reminded them that someone was thinking about them. With carpal tunnel and fibromyalgia, she still sent texts—a lot of them. And these were not mass or group texts. She sent individual texts, and made them personal. My mother often quoted Matthew 6:3 (NIV): "Never let your left hand know what your left right is doing." Well, she sure did a good job of that!

So, what's in your hand? How can you use it for the glory of God?

Spiritual Tips for Digital Evangelism

Tip #1: Move from online content to real life change.

Digital missionaries do more than share content. Your goal is not solely to impact people in the digital or virtual world, but to utilize online environments to impact the lives of people in the physical world. When I (Jeff) talk about "online to offline," I explain it by simply saying that "the gospel we learn in our online world has to influence our offline (physical) lives. If it's not impacting the physical, then all we're doing is creating consumers of a digital product."

You are using Web 2.0 and Web 3.0 technology not to just tell people about Jesus but to disciple them to the place where there is spiritual fruit in digital/metaverse space and also in physical space. That's the standard.

Tip #2: Show integrity in your online spiritual life.

The fruit of the Spirit should be consistent in your life—both online and off. For your digital ministry to be effective, you should be striving to grow in love, joy, peace, patience, kindness, goodness, faithfulness, gentleness, and self control digitally and physically. You must be consistent in your physical self as well as your digital/metaverse identity.

This is biblical, by the way. Pardon this slight paraphrasing of James 1:22: "Do not just be hearers of the word [online] and so deceive yourselves. Do what it says [in the physical world]."

Tip #3: Your digital reputation matters.

The way you behave in digital and metaverse spaces creates a reputation. If you were a business, I (John) would call it your brand, what people think of when they see your logo. Online, your name is like your logo. It's what people think of when they see your name. For instance, when you get a notification that Person A made a post, what do you anticipate that post to be about? Are you excited to read their post because you're thinking, "Oh, they always post insightful comments." Or are you thinking, "What are they trying to stir up now?" That's their reputation.

What do people think and feel when they see your name in a notification? Each one of us has a digital or metaverse reputation. The same way your reputation and what people think about you matters in person, it matters online as well. Remember Megan's story about praying for Lisa through TikTok? Lisa said, "You are the only person I know will pray with and for me." Why did Lisa expect that from Megan with such confidence? Because of how Megan consistently shows up like that on TikTok. That's what a reputation or brand is all about. It tells people what they can expect.

People choose the way they show up online, but they are not always aware of how others see them. Make sure that your profile, newsfeed, who you follow or like (because that's visible to others), and how you interact in chats, groups, rooms, etc. do not compromise your credibility when it comes to sharing the gospel.

Ministry Tips for Digital Evangelism

In addition to the great spiritual tips Jeff shared, I (John) would like to share some ministry tips.

Tip #1: Know when to shake off the digital dust.

If you run into someone who is aggressively opposed to Christianity, do not argue in a public chat. Arguing online is rarely fruitful. Be positive, end your involvement, and move on. You will know and feel the difference between someone who is asking questions for clarity and who wants to further understand your beliefs versus someone who is being antagonistic. Don't go there with them! "And if anyone will not receive you or listen to your words, shake off the dust from your feet when you leave that house or town" (Matthew 10:14, ESV). You have permission from Jesus to shake off the digital dust! Sometimes people can be very disrespectful. The longer you engage with them, the higher the probability of you making an emotionally charged response that, well, let's just say would not glorify God. A simple, "Looks like we just disagree, and that's OK. Blessings!" will do just fine.

Tip #2: Honor platform/group rules.

When in a social media subculture group, like a Facebook group, honor the rules of the platform and group. Don't be *that* person who has to be removed from the group or chat. For example, if you are in a group that says, "No politics or religion should be discussed" and you see a comment that you want to respond to with a spiritual response, send the person a direct message and let them know about it. That way, the person knows to look for your response, and the two of you can have that conversation privately.

Tip #3: Use direct messages.

As already mentioned a few times, take initiative to move conversations from public to private, and even from written communication to audio/visual.

Tip #4: Know when to stay on public chat.

There are times when I am in a public chat and I'm able to clarify a misunderstanding about a Christian idea. This is not an argument. Don't argue with people in chats. However, there are times when there is healthy discussion that can happen. This is an opportunity to bring truth or clarity to a person and the audience that is watching. For example, if someone directly attacks Christianity with, "All Christians/churches are mean," you can respond in a way that proves the opposite. This is a great opportunity, because for those who see Christians as mean, overbearing, manipulative, and frankly anything but loving, we get the chance to show the opposite. This is big when dealing with written communication because your positive witness for Christ will continue to be seen by others for days and even weeks later! If you are on platforms where there is live group communication, govern yourself well.

Tip #5: Follow-up is important.

Follow up with people God highlights to you. You don't need to be a Bible expert. You just need to be a digital listener. Whether you choose to send a direct message or post in the forum where your initial connection was made, people will feel valued. For example, "Hey, I remember you asked for prayer for a job interview. Just checking to see how it went." or "How is your mom doing? I've been praying for her."

Tip #6: Encouragement goes a long way.

Encouragement is a great way to open the door to relationship. It is another way to serve people. And let's face it, everyone could use more encouragement. No one is complaining that they are too encouraged. You can share simple statements like, "I saw that you are having a tough time right now. I just wanted to encourage you to keep your head up. You can make it! If you ever need to chat just let me know." A little encouragement goes a long way, especially nowadays.

Tip #7: Don't send that!

Over the past two years, I have typed and deleted more Facebook and Instagram posts than I have the many years before. There have been many comments I wanted to respond to or posts I wanted to initiate, but the Spirit said, "Don't send that!" Has that happened to you? Truthfully, there have been other times when I hit "Send" and moments later had to hit "Delete post." When that has happened, it's usually pretty obvious why the Spirit wanted me to refrain from posting.

But what I want you to consider is that you never know who is watching your posts. There are literally hundreds or thousands of people, depending on your followers, who are watching your content. We never know if it could close the door to the gospel or cause someone we do not know to stumble—but God knows. Trust his voice.

Become a Content Creator: Host Your Own Page or Channel

For those of you who have your own page or channel where you are sharing the gospel, I (John) want to encourage you to keep it up! Digital evangelism is filled with unknown impact to you as a content creator—unless someone lets you know what has happened on the other side of the screen. I love what York Moore shared about the impact of his TikTok page:

> One follower recently wrote these words: "Before I found your account, I didn't know anything about God. Now I am a Christian and am getting baptized!" I wake up to comments like this literally EVERY SINGLE DAY and the main reason is I provide distinct value in the form of constant teaching. Out of the 280,000 followers, I estimate 8,000 users are engaging with my content every day throughout the day.[55]

"And You Will Be My Witnesses"

Some of Jesus' last words to his followers in Acts 1:8 (ESV) are some of the last words I (John) want to leave you with in this book: "But you will receive power when the Holy Spirit has come upon you, and you will be my witnesses in Jerusalem and in all Judea and Samaria, and to the end of the earth."

"You will be my witnesses" . . . everywhere!

I grew up in a Black Baptist church. When the preacher made a strong point, he would often pause and ask, "Can I get a witness?" That was our cue to give a hearty "Amen" or shout! Every time I read this verse, I can't help but imagine Jesus literally asking, "Can I get a witness? Can I get someone to testify of my life-transforming power? Can I get a witness from someone I rescued from darkness? Can I get a witness from someone who knows what it's like to have their sins forgiven? Can I get a witness from someone who knows God as a Father? Can

I get a witness from someone who has experienced the power of God? Can I get someone to share how I completely turned their life around? Can someone tell a friend about the hope they have found in me?"

Jesus wants you to be his witness . . . everywhere, including online. Will you? If Jesus was giving this commission in 2023, I can imagine him saying, "And you will be my witnesses in news feeds, live streams, chat rooms, video chats, on webcams, and virtual worlds . . . and in future social technologies. Go into all the world. Go into all the online platforms, newsfeeds, and chat rooms, and compel them to come and be reconciled to our Father. Let them know how much I love them. Let them know that I have made a way to forgive them of *all* their sin so that we can be together for eternity."

ACTION STEPS: Moving B.L.E.S.S. Forward

Pray for Evangelism

» Acts 4:29-31 (NIV): "Enable your servants to speak your word with great boldness. Stretch out your hand to heal and perform miraculous signs and wonders through the name of your holy servant Jesus."

» Romans 10:1 (ESV): "My heart's desire and prayer to God for [the Israelites] is that they may be saved."

» Ephesians 6:19-20 (NIV): "Pray also for me, that whenever I speak, words may be given me so that I will fearlessly make known the mystery of the gospel, for which I am an ambassador in chains. Pray that I may the mystery of Christ, for which I am in chains. Pray that I may declare it fearlessly, as I should."

» Colossians 4:3-4 (NIV): "Pray for us, too, that God may open a door for our message, so that we may proclaim the mystery of Christ, for which I am in chains. Pray that I may proclaim it clearly, as I should."

» 2 Thessalonians 3:1 (NIV): "Pray for us that the message of the Lord may spread rapidly and be honored."

What's Next?

SO, DIGITAL MISSIONARY, YOU'VE LEARNED all about digital ministry. You've been taught all about B.L.E.S.S. You understand Beginning with Prayer, Listening, Engaging in Shared Experiences, Serving, and Sharing your Jesus Story. You've got the knowledge. Now, it's time to put the knowledge into practice. Start praying that the Holy Spirit will bring people to mind for you to build or restore relationships with. At Digital Church Network we've got an online course and community to encourage you in these B.L.E.S.S. steps.

Maybe some other people from your church would be interested in the challenge of becoming a digital missionary. What would happen if your church developed a team of digital missionaries, engaging on digital mission? There is so much potential in churches like yours if they will engage relationally.

If you're a church interested in expanding your ministry, or an individual just looking for some accountability, jump in our community at http://digitalchurch.network/missionary to check out our online course or to find more information.

ACKNOWLEDGEMENTS

Dr. John Harris

To my wife, Jeanine, and my three girls, Elise, Sydney, Grace, thank you for your love, support, encouragement, and inspiration in my life and ministry.

To my family and friends, thank you for your years of prayers and support.

To my Hearts Like His family, thank you for your prayers, love, and support as I learned the ropes as a digital pastor.

To Jeff Reed, my co-author, thanks for connecting me with Exponential and for sharing your platform with me.

Dave Ferguson and Exponential, thank you for your trust in me to write this book. Thanks for inviting me to be a part of your work to help churches around the world multiply at every level.

Jeff Reed

To Amy, Caitlyn, and CJ, thank you for your love and support . . . and for thinking that my nerdy life is cool.

To Dr. John, thank you for challenging people, myself included, in evangelism and transformative disciple-making in digital spaces.

To my Digital Church Network FAM and my THECHURCH.DIGITAL friends, humbly, thank you for thinking differently.

To Dave Ferguson and the Exponential, Leadership Network, and NewThing Network ecosystem, your support of digital and metaverse ministry has been incredible. Thank you for what you've done for me personally and for others who are doing ministry in these unique spaces.

To those I've encountered on my 20+ year digital ministry journey, stay tuned. We're just getting started.

ENDNOTES

1 Jeremy Myers, "65 Million American Adults Have Left the Church?," Redeeming God, October 2015 (approx.), https://redeeminggod.com/adults-have-left-the-church/.

2 Dave Ferguson, *B.L.E.S.S.: 5 Everyday Ways to Love Your Neighbor and Change the World* (Washington, DC: Salem Books, 2001), chap. 1, Kindle.

3 "Global Social Media Statistics," DataReportal, October 2022, https://datareportal.com/social-media-users.

4 Brandon Clay, "Top 101 Theology Questions in 2022 (According to Google)," TruthStory, February 20, 2022, https://truthstory.org/blog/top-101-theological-questions/.

5 Jeffrey Jones: "Confidence in U.S. Institutions Down; Average at New Low," Gallup, July 5, 2022, https://news.gallup.com/poll/394283/confidence-institutions-down-average-new-low.aspx.

6 Aaron Earls, "Americans' Trust of Pastors Hovers Near All-Time Low," Lifeway Research, January 22, 2021, https://research.lifeway.com/2021/01/22/americans-trust-of-pastors-hovers-near-all-time-low/.

7 "2 in 5 Churchgoers Are Open to Inviting Others to Digital Church Services," Barna, December 2, 2020, https://www.barna.com/research/digital-service-invitation/.

8 Jeff Reed, "Megaphone or Telephone? What Billy Graham Taught Me About Digital Communication," THECHURCH.DIGITAL, August 1, 2018, https://be.thechurch.digital/blog/megaphone-or-telephone-what-billy-graham-taught-me-about-digital-communication.

9 Jeff Reed, "Roundtable: Dave Ferguson, Digital Missionaries and the Power of Shared Experiences," Digital Church Network, May 2022, video, 1:01, https://www.youtube.com/watch?v=K6Hr1r0OFrA.

10 Ibid.

11 *The Chronicles of Narnia: The Lion, the Witch, and the Wardrobe,* directed by Andrew Anderson (Burbank, CA: Walt Disney Pictures and Walden Media, 2005).

12 C. S. Lewis, *The Lion, the Witch, and the Wardrobe* (New York, NY: Scholastic Inc., 1950), 67.

13 Angela Craig, *Online Jesus: A Guide to Community, Discipleship, and Care Online* (independently published, 2020), chap. 22, Kindle.

14 "The Church That Meets on WhatsApp," Jesus.net, accessed January 2, 2023, https://jesus.net/the-church-that-meets-on-whatsapp/.

15 Taryn Hayes, "Innovative Evangelism: Bringing the Gospel to TikTok," The Gospel Coalition, July 10, 2020, https://au.thegospelcoalition.org/article/innovative-evangelism-bringing-the-gospel-to-tiktok/.

16 Christopher Hutton, "God Mode Activated: Meet the Gamers Bringing Jesus to Twitch," OneZero, November 17, 2020, https://onezero.medium.com/god-mode-activated-meet-the-gamers-bringing-jesus-to-twitch-3fb68227f57e.

17 York Moore, "TikTok Evangelism," Campus Ministry Today, June 8, 2022, https://campusministry.org/article/tiktok-evangelism.

18 Jeffrey Jones, "U.S. Church Membership Falls Below Majority for First Time," Gallup, March 29, 2021, https://news.gallup.com/poll/341963/church-membership-falls-below-majority-first-time.aspx.

19 Moore, "TikTok Evangelism."

20 Chris Kolmar, "25 Amazing Virtual Reality Statistics [2022]: The Future of VR + AR," Zippia: The Carrier

Expert, October 9, 2022, https://www.zippia.com/advice/virtual-reality-statistics/.

21 Internet search, "How many people live in the US?", Google.com, accessed January 2, 2023, https://www.google.com/search?q=how+many+people+live+in+the+us.

22 Jonathan Sprowl, "Ministry in the Metaverse," Outreach Magazine, November 23, 2021, https://outreachmagazine.com/features/leadership/69751-ministry-in-the-metaverse.html.

23 "How Much Time Does the Average Person Spend on Social Media? (2012-2022)," Oblero, accessed January 2, 2023, https://www.oberlo.com/statistics/how-much-time-does-the-average-person-spend-on-social-media.

24 "META: Company Information," META website, accessed January 2, 2023, https://about.meta.com/company-info/.

25 Randy Larson and Jeff Reed presented these Barna statistics in "Hope Among Sobering Statistics" (slide presentation produced by GLOO, November 2022), slide #16.

26 Ibid.

27 Chris Coursey, *Transforming Fellowship: 19 Brain Skills That Build Joyful Community* (Holland: Thrive Today, 2016), 45.

28 J. D. Greear, *Gospel Revolution* (Nashville: B&H Publishing Group, 2011), 21.

29 "Why You Shouldn't 'Preach the Gospel at All Times and Use Words When Necessary,'" Zondervan Academic, June 27, 2016, https://zondervanacademic.com/blog/why-you-shouldnt-preach-the-gospel-at-all-times-and-use-words-when-necessary.

30 Glenn Stanton, "FactChecker: Misquoting Francis of Assisi," The Gospel Coalition, June 10, 2012, https://www.thegospelcoalition.org/article/factchecker-misquoting-francis-of-assisi/.

31 Mark Galli, "Speak the Gospel," Christianity Today, May 21, 2009, https://www.christianitytoday.com/ct/2009/mayweb-only/120-42.0.html.

32 Ibid.

33 Ferguson, B.L.E.S.S., chap. 3 (Kindle).

34 "Charles Finney: Father of American Revivalism," Christianity Today, 1988, https://www.christianitytoday.com/history/people/evangelistsandapologists/charles-finney.html.

35 J Paul Reno, "Daniel Nash 1775-1831 – Prayer Warrior for Charles Finney," Hope Faith Prayer: The Mechanics of Faith, accessed January 2, 2023, https://www.hopefaithprayer.com/prayer-warrior-charles-finney/.

36 Ibid.

37 Werner Geyser, "32 Facebook Statistics – Revenue, Users (+ Everything You Need to Know in 2023)," Influencer MarketingHub, December 23, 2022, https://influencermarketinghub.com/facebook-statistics/.

38 Ferguson, *B.L.E.S.S.*, chap. 4 (Kindle).

39 Craig, *Online Jesus*, chap. 1 (Kindle).

40 Ben Lutkevich, "social listening (social media listening)," TechTarget, February 2022, https://www.techtarget.com/searchcustomerexperience/definition/social-media-listening.

41 Ferguson, *B.L.E.S.S.*, chap. 4 (Kindle).

42 Ibid.

43 Ibid.

44 Francis Schaeffer, quoted by Jerram Barrs in Francis Schaeffer: His Apologetics, chapter 2 of *Francis Schaeffer: A Mind and Heart for God*, http://francisschaefferstudies.org/?feed=html/social-feed/date/2017-04-18.

45 Anthony English, "The Three E's of Evangelism, Modern Reformation," April 29, 2022, https://modernreformation.org/resource-library/web-exclusive-articles/the-mod-the-three-es-of-evangelism.

46 Ferguson, B.L.E.S.S., chap. 4 (Kindle).

47 Ferguson, *B.L.E.S.S.*, chap. 5 (Kindle).

48 https://www.nytimes.com/games/wordle/index.html.

49 Ferguson, *B.L.E.S.S.*, chap. 6 (Kindle).

50 Ibid.

51 "God Speaks Through His Activity," Blackaby Ministries International, October 25, 2022, https://blackaby.org/god-speaks-through-his-activity/.

52 Ferguson, *B.L.E.S.S.*, chap. 7 (Kindle).

53 Ibid.

54 Seth Godin, "Sample Size," *Akimbo: A Podcast from Seth Godin*, February 12, 2020, https://play.acast.com/s/akimbo/samplesize.

55 Moore, "TikTok Evangelism."

ABOUT THE AUTHORS

Dr. John Harris

In over 30 years of ministry, Dr. John Harris has served as a chaplain in the United States Air Force (with tours in Iraq and Afghanistan), a Christian education pastor, a teaching pastor, a small groups pastor, and an executive pastor. He has a passion to see every expression of the Church grow and believers being equipped to fulfill the mission of Christ.

In January of 2021, Dr. John and his wife launched Hearts Like His, an online discipleship community on Facebook. Their focus on digital discipleship now reaches people across the United States and in several other nations. He is a coach for digital church planters and a digital ministry consultant for churches.

Dr. John also serves as a leadership coach for pastors, executives, and senior managers. He facilitates workshops on emotional intelligence to increase leadership capacity, decrease team conflict, build teamwork and collaboration, and ultimately maximize organizational impact.

He holds a doctorate in transformational discipleship from Gateway Seminary. John lives in Roseville, California, with his amazing wife, Jeanine, and their three beautiful daughters, Elise, Sydney, and Grace.

Jeff Reed

In June 2000, Jeff Reed led his first online Bible study, taking 75 people from around the world through the book of James using a text-based system called Ultimate BB. He was doing digital ministry way before it was cool.

Since founding THECHURCH.DIGITAL in 2018, Jeff's passions have evolved into helping churches (and individuals too!) find their calling through digital discipleship, releasing people on digital mission, and planting multiplying digital churches. This pursuit is realized through DigitalChurch.Network, an organic, decentralized network for digital expressions of church and digital missionaries globally.

As author of *VR & the Metaverse Church* and now *Sharing Jesus Online*, Jeff champions digital and metaverse church planting while working with Leadership Network, NewThing Network, Exponential, He Gets Us, Media Impact International, and other globally facing, multiplication-friendly, gospel-centric organizations.

Jeff married his high school sweetheart, Amy, and has two kids (Caitlyn and CJ) and two dogs. They live in Miami, Florida.

THECHURCH.DIGITAL
WE HELP CHURCHES GO DIGITAL
More than just broadcasting church services, we help churches find their digital purpose grounded in missional focus and disciple-making.
SERVICES
CHURCH ASSESSMENT
We'll walk your team through our exclusive assessment, helping you find the opportunities and obstacles toward digital discipleship.
ONGOING CONSULTING
Sometimes churches need more than the map; they need someone to join them on the journey. We will walk your road with you, helping you along the way.
FREE CONSULTATION
It all starts with a free conversation with one of our digital experts. Start your journey toward understanding digital discipleship TODAY!
ABOUT US
With (combined) over 50 years of digital ministry experience, THECHURCH.DIGITAL team has an established track record of developing healthy, successful digital ministry for churches.
WHY CHOOSE US ?
Jeff, Dr. John, and the rest of the team are excited to help your church discover your "why" of digital ministry and help it align with what your church is doing physically.
MEET THE TEAM AND DISCOVER DIGITAL PURPOSE AT YOURCHURCH.DIGITAL

Made in the USA
Monee, IL
14 July 2026

57205712R00095